Arthur Haberman
Europe, 1859

Arthur Haberman

Europe, 1859

In the Ebb and Flow of Modernity

DE GRUYTER
OLDENBOURG

ISBN 978-3-11-153673-6
e-ISBN (PDF) 978-3-11-079308-6
e-ISBN (EPUB) 978-3-11-079315-4

Library of Congress Control Number: 2022944043

Bibliographic information published by the Deutsche Nationalbibliothek
The Deutsche Nationalbibliothek lists this publication in the Deutsche Nationalbibliografie; detailed bibliographic data are available on the internet at http://dnb.dnb.de.

© 2024 Walter de Gruyter GmbH, Berlin/Boston
This volume is text- and page-identical with the hardback published in 2023.
Cover image: Map, Heritage Images / Mithra – Index / akg-images. Background, jessicahyde / iStock / Getty Images Plus

www.degruyter.com

To Jan

"By 'modernity,' I mean the ephemeral, the fugitive, the contingent, the half of art whose other half is the eternal and the immutable... For the perfect *flâneur*, for the passionate spectator, it is an immense joy to set up house in the heart of the multitude, amid the ebb and flow of movement, in the midst of the fugitive and the infinite."

Charles Baudelaire

Acknowledgements

I am very grateful for the valuable assistance I received from Fran Cohen, Martin Sable and Adrian Shubert. Jan Rehner read carefully, asked incisive questions and, as always, supported me in a thousand ways.

The team at De Gruyter was excellent. Rabea Rittgerodt, Benedikt Krüger, and Matthias Wand were very supportive and helpful.

Contents

List of illustrations —— XIII

Introduction —— 1

Prologue – On Being Modern —— 4

Chapter I
Baudelaire and Manet: The Experience of Modernity —— 8

Chapter II
Giuseppe Mazzini: Prophet of Nationalism —— 41

Chapter III
Ivan Turgenev: Russia on the Eve —— 58

Chapter IV
Jacob Burckhardt: Inventing the Renaissance —— 78

Chapter V
John Stuart Mill: Liberty and Modernity —— 93

Chapter VI
The New City: Manchester, Paris, and Barcelona —— 112

Chapter VII
Karl Marx and Friedrich Engels: Understanding Industrial Society —— 143

Chapter VIII
The Anti-Moderns —— 169

Chapter IX
Charles Darwin: The Mystery of Mysteries —— 187

Epilogue – Reflections —— 209

Index —— 221

List of illustrations

Fig. 1: Gustave Caillebotte, *Young Man at His Window*, 1876, Source: https://www.getty.edu/art/collection/object/10A2Y1
Fig. 2: Gustave Caillebotte, *Le Pont de l'Europe*, 1876, Source: akg-images
Fig. 3: Pierre Auguste Renoir, *La Loge*, 1874, Source: Heritage Images / Fine Art Images / akg-images
Fig. 4: Edouard Manet, *The Absinthe Drinker*, 1859, Source: akg-images
Fig. 5: Honoré Daumier, *The Exposition of 1859*, Le Charivari, April 18, 1859, Source: www.daumier.org
Fig. 6: Edouard Manet, *Music in the Tuileries*, 1862, Source: The National Gallery, London / akg
Fig. 7: Claude Monet, *Impression, Sunrise*, 1872, Source: Heritage Images / Fine Art Images / akg-images
Fig. 8: Edouard Manet, *Le déjeuner sur l'herbe*, 1863, Source: akg-images / Laurent Lecat
Fig. 9: Claude Monet, *Le Pont de l'Europe – Gare Saint Lazare à Paris*, 1877, akg-images
Fig. 10: Edouard Manet, *Gare Saint-Lazare* (also called *Chemin de Fer* or *The Railway*), 1872, Source: akg-images
Fig. 11: Edouard Manet, *Bar at the Folies-Bergère*, 1882, Source: Heritage-Images / The Print Collector / akg-images
Fig. 12: *Peterloo Massacre*, Print published by Richard Carlile, October 1, 1819, Source: akg-images / WHA / World History Archive
Fig. 13: Honoré Daumier, *Tenants and Landlords*, Le Charivari, February 21, 1854, Source: www.daumier.org
Fig. 14: Ildefonso Cerda, *Map of the project of expansion of the city of Barcelona* (Cerdà's 1859 plan), Source: Heritage Images / Mithra – Index / akg-images
Fig. 15: Honoré Daumier, *Members of the Charitable Society of December 10 Demonstrating their Philanthropic Activities*, October 1850, Source: www.daumier.org
Fig. 16: Honoré Daumier, *Proudhon*, 1850, Source: Bibliothèque municipale de Besançon

Introduction

When does modernity begin in Europe? The question has had a variety of answers.

Some place it in the Renaissance, with the revival of antiquity, the growth of the secular city-state, the value placed in the exceptional individual, and the beginnings of the discovery of the larger world.

Others find it in the Enlightenment, with the growth of the concept of human rights, the importance of the new science in relating to the world, the belief in progress, and the challenge to the feudal world.

Many go to the French Revolution. 1789 marks the political transformation into modernity—the politics of the street, the reconstituting of law and power, the ideas of human rights as fundamental to civic culture, and the end of the *ancien régime*.

Others go to the Industrial Revolution, dated variously from 1780 to 1820. This of course transformed how we labor and how wealth is created and understood. It marked the urbanization of life and the class structure that endures until today.

Another group simply goes to both, the "dual revolutions" discussed by Hobsbawm and others.

There will likely never be agreement on the concept of the modern, because it will be defined in the West in a variety of ways.

But when was it confronted, thought about, marked as something that requires new modes of understanding and new ways of dealing with the quality of life, the economics of capitalism, the rise of the new bourgeoisie to power, the politics of liberalism and socialism? This does not occur on a large scale until even later than the beginnings of the dual revolution, until poets, novelists, artists, philosophers, and scientists confront the new reality.

In 1859, Charles Baudelaire is writing the poetry and criticism of the new urban cultural and social world which would make him described by a number of historians as the first modern. Indeed, it is he who coined the term "modernity." In the east, Ivan Turgenev with *On the Eve* begins reflections about Russia and modernity which would result in his next novel, set in 1859, *Fathers and Sons*. The latter still resonates today. In Switzerland, Jacob Burckhardt is inventing the Renaissance as a means of understanding what is happening in his own time. Indeed, we never talked about a Renaissance until Burckhardt published his *The Civilization of the Renaissance in Italy* in 1860, something he wrote in order to better understand his own times. In the West, several important and central works of European culture are being written in England by both British writers and exiles. Marx is researching *Das Capital* and writing *A Contribution to the Critique of Political Economy*. Mazzini is writing his major work on modern nationalism, *The Duties of Man*, just as Italy is beginning its decade of unification and the European map is beginning a period of extraordinary change. John Stuart Mill published his *On Liberty* in February 1859, still the work that is the modern ground of democratic ideas dealing with the relationship between liberty and authority. And in November 1859 one of the dozen or so most influential works of all of European history and science, one that shattered many pre-modern concepts, *The Origin of Species*, was published by Charles Darwin.

It was in 1859 that modernity, the world as we now know it, gets confronted and encountered. As a result concepts and ideas we still use, then new, get thought about and become part of the public discourse. From this point on, the dialogue is forever transformed.

What were the main issues?

How do we understand and respond to new modes of living resulting from the new political discourse, urban life, and the new economic reality?

Is it necessary to write, paint, and do music in different ways in order to capture the modern experience and reality? Do we need a new language and philosophy to comprehend what is happening and what to do about it?

How do we understand identity and aspirations now that the old theology and the central position of churches is challenged?

What is to be made of the new industrial society? What is just? What ought to happen with the new wealth and new classes?

Has nationalism become the new way of organizing our political and civic culture? What are the consequences of the new national identity?

How do we understand human nature? Are we those who belong to a people? Are we members of a class? What does it mean to have consciousness? Are we descended from the animal kingdom?

The thinkers who were prominent at the time were, in a full sense, public intellectuals. Their works were read, debated, applauded, feared, defended, and scorned in the public fora, what philosophers sometimes called the marketplace. Moreover, they were not ivory tower intellectuals. One of the keys to modernity is the union between thought and action. It was Marx who stated: "The philosophers have only *interpreted* the world in various ways; the point, however, is to *change* it." They are of and in the world. And, indeed, they did both help us to understand the new modernity and they contributed profoundly to social, political, and cultural transformation.

This study is centered on two matters: first, at its core is the significance of the year 1859 in the intellectual life of the West, as a number of important works,

works that still matter, were published, beginning with *On Liberty* in February and going forward to *The Origin of Species* in late November. It also has as a major leitmotif the idea that if there is a single year in which important thinkers and public intellectuals encountered modernity and reflected on it, that year is 1859, a *wunderjahr*, a wonder-year, in the cultural history of the European continent.

In organizing this work and planning who to discuss, I encountered some important matters related to the Other and to gender. Simply, the main figures are all white males who researched and wrote and painted at the time. My dilemma was that there were few female authors or painters who did work which was as important as, say, Mill's *On Liberty* or Baudelaire's prose poems or Darwin's *Origin of Species* in 1859. I consulted readers and scholars and, while some mentioned George Eliot and/or George Sand, neither wrote breakthrough major works around our *wunderjahr*. Moreover, in looking into Black culture in Europe at the time, the only name mentioned was that of Alexandre Dumas, the offspring of a white European and a Caribbean slave, whose works also did not fit. These omissions trouble me because in an earlier work of mine centered around the year 1930, I was easily able to include Virginia Woolf and other women, as well a chapter devoted to Black culture in Paris and to the Nardal sisters, Paulette and Jane.

I have attempted to address this matter as one related to the experience of the Other, including women in modernity. Hence, there is a part of the first chapter which explains that women did experience modernity, but much later than in the mid-nineteenth century. As well, when appropriate I give attention to the contributions of women such as Mary Cassatt and Elizabeth Gaskell. As Mill himself pointed out in his *The Subjection of Women* (1869), Western culture and society were narrow and short-sighted on matters of gender at the time. As Mill remarked: "We have had the morality of submission, and the morality of chivalry and generosity; the time is now come for the morality of justice."

Prologue – On Being Modern

Thinking about being modern is something relatively recent in European cultural history. Valuing it is even more so.

The two major mythic structures of the West tell us to view the present as a kind of regress, a descent from a perfect moment or a golden age. The Bible puts the Garden of Eden at the beginning of time, even outside of history, because there is no change. Only after the expulsion does human history begin, as both told in the text of Genesis or in the more modern recounting of the myth by Milton in *Paradise Lost*.

The Greco-Roman myth also puts the golden age at the beginning of time, as told by Ovid and others. It was a perfect time, claimed Ovid, "that cherished/Of its own will, justice and right... Spring was forever...." This was succeeded by the ages of silver and bronze. Then his own time, an age of iron which "let loose all evil." Violence was ordinary. The earth was abused, and war and blood were a large part of human experience.

Hence, being in the present was a kind of lament, a statement about humanity's base nature, an experience of suffering and evil, and a longing for what was lost.

That sentiment was echoed in the fourteenth century by one of the early humanists, Francesco Petrarch. He longed to be in the classical world and wrote letters to his heroes of ancient times. He told Cicero he was guided by him and through following him he became a fine writer. He wrote to Seneca, telling him of the pleasure he had in speaking with him. He admitted to Livy that "I should wish (if it were permitted from on high) either that I had been born in thine age or thou in ours." He longed to live in ancient times, for his own time, Petrarch believed, was much inferior to that of the Classical era.

By the fifteenth century some expressed positive things about their present, though most in the Italian Renaissance still looked to the classics for guidance and sought to revive an earlier, better time. Still, Vespasiano da Bisticci could list the qualities of illustrious men of his time and include such virtues as knowledge, liberality, the love of books and the collection of a library, the support of scholars, and the composition of excellent verse. Pico de Mirandola, a controversial figure, could praise the possibilities of being human and to be able to choose a good life. "If rational," he wrote, "[a man] will grow into a heavenly being. If intellectual, he will grow into an angel and the son of God... Who would not admire this our chameleon."

As we shall see, the concept of a Renaissance which is modern is itself a creation of the nineteenth century, around our year of 1859. However modern Renaissance philosophers and scholars are seen by us to have been, they still found themselves trying to emulate and catch up to the ancients. An example would be the excellent Machiavelli, whose political thinking we find modern, but whose examples are almost all classical.

The battle between the ancients and moderns, as many have termed it, was lost to the ancients until at least the eighteenth century. Indeed, most from the fourteenth to the eighteenth century were happy to lose that battle, grateful that after many centuries they now had access to ancient knowledge and wisdom.

It is in the last half of the eighteenth century that some important thinkers began to believe that being modern was superior to being in the ancient world. In 1784 Immanuel Kant argued that he was now living in the age of Enlightenment. There was much to do, he argued, before he would state that he was in an enlightened time, but it was now possible with the application of reason. We could now "dare to be wise" and move away from our "self-incurred immaturity."

Others took up Kant's program, most notably the philosopher the Marquis de Condorcet. He wrote one of the great testaments to the idea of progress in 1793 and 1794 (while in hiding from the Jacobins during the French Revolution), in his *Sketch for a Historical Picture of the Progress of the Human Mind*. He said:

> Such is the object of the work I have undertaken; the result of which will be to show, from reasoning and from facts, that no bounds have been fixed to the improvement of the human faculties; that the perfectibility of man is absolutely indefinite; that the progress of this perfectibility, henceforth above the control of every power that would impede it, has no other limit than the duration of the globe upon which nature has placed us. The course of this progress may doubtless be more or less rapid, but it can never be retrograde; at least while the earth retains its situation in the system of the universe, and the laws of this system shall neither effect upon the globe a general overthrow, nor introduce such changes as would no longer permit the human race to preserve and exercise therein the same faculties, and find the same resources.

Condorcet talked about emancipating humanity "from its chains"[1] and, among other matters, discussed ideas of universal education, the equality of the sexes, the further development of the sciences, a greater equality of wealth, the amelioration of the inequality of condition, and progress in the arts and industry. He predicted a humanity "restored to its rights, delivered from oppression, and proceeding with rapid strides in the path of happiness."

The most notable and influential "modern" in the first half of the nineteenth century was Claude-Henri de Rouvroy, Comte de Saint-Simon, commonly now

---

1 A nod to Rousseau and an anticipation of Marx.

known as Henri de Saint-Simon (1760–1825). He lived through the French Revolution and the Napoleonic regime and became one of the major political and economic commentators of his time in the last decade or so of his life.

In October 1814, Saint-Simon had a radical solution to the dilemma of how Europe should be organized after the defeat of Napoleon, one far different from that taken ten months later at the Congress of Vienna. In "The Reorganization of the European Community," co-authored with his pupil Augustin Thierry, he became among the first to propose a unified Europe in what he termed "a single body politic." We are now enlightened, he argued, and hence we need to put the social and political systems in better correspondence with the level of Enlightenment. Common interest must come before national interest: "[T]hen evils will begin to lessen, troubles abate, wars die out."

Saint-Simon ended with one of the great statements supporting the idea of progress and the importance of the modern:

> Poetic imagination has put the Golden Age in the cradle of the human race, amid the ignorance and brutishness of primitive times; it is rather the Iron Age which should be put there. The Golden Age of the human race is not behind us but before us; it lies in the perfection of the social order. Our ancestors never saw it; our children will one day arrive there; it is for us to clear the way.

In 1819, Saint-Simon suggested to his readers that they join him in a mental exercise. Suppose, he said, that France lost its best physicists, mathematicians, engineers, artisans, professionals and artists. Then suppose that France lost the king's brother, other nobles, cardinals and bishops, officers of the royal household, judges, and those who are rich and live like the nobility. The argument was simple. Anyone can replace the latter; no one can replace the great physicists and others. We have, he stated, a system in which the ignorant and the idle rule over those who are creative, industrious, and contribute to the betterment of society.

There came to be in the next two generations a number of Saint-Simonians, people who believed that it was important to modernize and to transform society, sometimes with the state leading the way. The most prominent of all these came to be Louis Napoleon Bonaparte, later known as the emperor Napoleon III, who ruled France in the 1850s and 1860s, and who we shall again encounter in this study.

It is impossible to discuss major figures who reflected on the modern without mentioning Karl Marx (1818–1883), who as early as 1844 grappled with understanding the economic and social issues surrounding the new industrial revolution. He, too, believed in progress, even, as with Condorcet, its inevitability. He, too, like Saint-Simon, warned us never to consider political and social matters independent of the economic substructure of any area.

It is difficult to assign a single person the credit for the introduction of the idea of the consciousness of being modern. Certainly, from the late eighteenth century, Kant, Condorcet, Saint-Simon, and Marx deserve serious consideration, as do others. And Burckhardt redefined the world of Italy in the fourteenth and fifteenth centuries to give us a sense that these people were not unlike what we have become, even if it did not last.

However, two mid-century nineteenth century figures must also be put in the first rank. Charles Baudelaire (1821–1867) is often said to be the "first modern," because not only did he invent the term modernity, but was the first poet and critic to deal deeply with the experience of modernity. His friend, Édouard Manet (1832–1883), was as an artist the painterly equivalent of Baudelaire. Manet not only painted his time, but had a deep consciousness of its uniqueness and what it meant to experience it. Both men explored modern consciousness in the 1850s and 1860s. Both gave us the first sense of what it meant as an individual to be modern and to critique being modern at the same time.

Chapter I
Baudelaire and Manet: The Experience of Modernity

Between 1857 and 1867 Charles Baudelaire composed the 50 prose poems that were not published together as *Le Spleen de Paris* until 1869, two years after his death. The poems are Baudelaire's testament to the experience of modern life, to both the joys and anguish of being conscious and living in a great city in the modern age.[2]

Baudelaire became notorious in 1857. In that year he published his *Les Fleurs du Mal*, a volume of poems which did not mention God but often referred to the satanic and to evil. In them, he wanted to find beauty in evil, in the contradictions and ambiguities of life in the mid-nineteenth century, in the grittiness of human existence. As well, he experimented with poetic form and aesthetics, seeking to find in the imagination and in symbolism a new way of understanding and communicating. *Les Fleurs du Mal* made his reputation, this Parisian dandy who was already an art critic of some stature, the romantic Delacroix his favourite painter.

1857 is also important in French literary history for two public trials, the first related to Flaubert's *Madame Bovary* and the second to Baudelaire's poetry book. Both were condemned by the public prosecutor for being "*delit d'outrage à la morale publique*," offensive to public morality. Flaubert won his case early in the year. Baudelaire was ordered to censor six poems from the collection and to pay a fine of 300 francs. Most later editions of *Les Fleur du Mal* restored the excised poems.

Baudelaire invented the term "modernity" in his 1859–1860 essay, published in 1863, *The Painter of Modern Life*. In it he celebrates art that deals with the contemporary experience, at that time judged by him to be the work of the artist Constantin Guys (1802–1892). He praises Guys for seeking "to extract from fashion whatever element it may contain of poetry within history, to distil the eternal

---

[2] Note: unless otherwise indicated [by the initials AH] the translations of poems from *Les Fleurs du Mal* are from *Charles Baudelaire, The Flowers of Evil*, selected and ed. M. and J. Mathews, revised edition (New York: New Directions, 1963). The poems from *Le Spleen de Paris* are from Charles Baudelaire, *Paris Spleen and La Fanfarlo*, trans. Raymond N. MacKenzie (Indianapolis: Hackett, 2008).

from the transitory."³ Guys and artists interested in modern life are contrasted with David and other nineteenth century artists who choose subjects related to the history and myths of the ancient world. Now the modern was to be the subject of study.

Baudelaire's concern with finding how to understand the present recalls Marx's statement in his 1852 essay, *The Eighteenth Brumaire of Louis Bonaparte*, a discussion of the failure of the revolution of 1848 in France:

> Men make their own history, but they do not make it as they please; they do not make it under self-selected circumstances, but under circumstances existing already, given and transmitted from the past. The tradition of all dead generations weighs like a nightmare on the brains of the living. And just as they seem to be occupied with revolutionizing themselves and things, creating something that did not exist before, precisely in such epochs of revolutionary crisis they anxiously conjure up the spirits of the past to their service, borrowing from them names, battle slogans, and costumes in order to present this new scene in world history in time-honored disguise and borrowed language.⁴

For Baudelaire and others in 1859 there is a new world, one in which Becoming is reality and Being doesn't exist. We cannot understand the world in which we live using the categories of past times. This is not only true for Baudelaire. Others in the 1850's—Marx, Darwin, Mazzini, Mill—searched for new ways of dealing with the uniqueness of the modern world.

"By 'modernity,'" Baudelaire states, "I mean the ephemeral, the fugitive, the contingent, the half of art whose other half is the eternal and the immutable." Modernity is about living with constant transformation, in a whirl, especially for those who live in large urban areas. Everything is, he says, "transitory," all filled with "metamorphoses." Hence, a new set of critical and artistic values and concerns are needed to deal with this phenomenon.

A new class is rising, and they are now to be the subject of art and literature. Earlier, in his essay *The Salon of 1846*, Baudelaire became one of the first in the world of art to recognize the importance of the new bourgeoisie.

He dedicates the essay to the bourgeoisie. "You are the majority, in number and intelligence," he tells them.⁵ You now have power and you now are those who govern. Something new is happening, and, as clear as is Marx, Baudelaire also understands the transformation into modernity in class terms.

---

3 *The Painter of Modern Life and Other Essays*, trans. and ed. by Jonathan Mayne (London: Phaidon Press 1964), 12.
4 Robert C. Tucker, *The Marx-Engels Reader*, (New York, Norton, 1978), 595.
5 *Baudelaire: Selected Writings on Art and Artists*, trans. P.E. Charvet (Penguin, 1972), 47.

Moreover, the new bourgeoisie have begun to create new institutions, most notably in the art world. You have created museums, he noted, established collections, and made art accessible in a way that could not have happened before the July Revolution of 1830. Baudelaire further acknowledges the role played by the bourgeoisie in industry and economics. For him, the period of the Orleanist monarchy of Louis-Philippe (1830–1848) is indeed one which acknowledges the changes wrought from the earlier major revolution of 1789, that which transformed France and Europe. You have wealth, he says, but you also need poetry and art. It is not clear whether Baudelaire read St. Simon; yet his love of the modern and its new forms are not dissimilar.

He goes on to assert that the modern has many of the qualities that heretofore attracted artists and poets to the classical world. Modern life can be both heroic and epic: "...[T]he heroism of modern life surrounds and presses upon us."[6] For Baudelaire, we need not any longer go to the classical world for examples of great love, great suffering, great tragedy or great epics. "Our age is no less rich than ancient times in sublime themes," he tells his bourgeois readers.[7] Each age has its own variety of beauty, and we have ours.

Hence, Baudelaire is here echoing some of the themes of Romanticism. He identified deeply with the Romantic tradition, though he did admire some Realists—as soon as you think you have categorized someone like Baudelaire, he, like Nietzsche, slips away from you. Many Romantics, in England, France, and Germany, insisted earlier on the importance and mythic quality of their present, including Wordsworth in his poetry, Gericault and Goya in their art—most especially in Gericault's *The Raft of the Medusa* and in Goya's paintings of the Spanish uprising of May 2, 1808—and Friedrich in his understanding of the relationship between humans and nature.

Baudelaire goes further. For him, the ordinary has in it the extraordinary. Life in the bourgeois city—Paris is his home—is full of splendour, rich in possibilities and subjects, full of the marvellous and the tragic. You needn't go back to Homer's world. We, now, he claims, have our own aesthetic and richness, different from but equal to ancient lives: "The specific element of each type of beauty comes from the passions, and just as we have our particular passions, so we have our own type of beauty."[8]

Thus Baudelaire gives himself and his fellow poets and artists a new task. Tell what it means to be modern. For him, it is not something that is necessarily

---

6 Charles Baudelaire, *The Salon of 1845*, xx.
7 Charles Baudelaire, *The Salon of 1846*, 104.
8 Op. cit., 105.

consistent and it is certainly not something that is necessarily rational. Human nature contains within it contradictory emotions and paradoxical feelings. It is subject to metamorphosis at any time. His Paris is wonderful and terrible, sublime and pastoral, beautiful and ugly, all at the same moment.

As well, Baudelaire is not at all a believer in the idea of progress. He is among the first in the industrial era to challenge the notion that the development of industry and the new steam, electric, and gas power are indications of progress. Rather, he defines progress as moral rather than as something in the physical world. He attacks the conflation of the normative—the increase in industrial capacity and wealth, for example—with the valuative—progress in the distribution of food or in the arts. He certainly saw early that an increase in material wealth could not in itself signal any sort of progress in a valuative sense.

Baudelaire also abandoned the teleological assumption behind much of the philosophy underlying the idea of progress. To Saint-Simon and others, including Marx, progress was inevitable. This is absurd, claimed Baudelaire. There is no guaranteed movement forward. It is, he said, foolishness.

It is the modern city in countries that are industrializing that is the landscape of modernity for Baudelaire. And it is the *flâneur* who is the observer and recorder of modern times.

Paris had a population of 546,856 in 1801; by 1861 it grew to 1,696,141. London had 1,096,784 people in 1801, fast becoming, by 1815, the largest city in the world. By 1860 it had tripled to 3,188,485. Manchester, the example to the world of the new industrial city, had 89,000 in 1800; in 1851 it recorded about 400,000.

There was a double push in the large industrial towns. That is where the jobs came to be, and people migrated to the cities from the countryside. In addition, the nineteenth century witnessed the beginnings of a demographic revolution that persists today. After centuries of stability or slow growth the curve bent radically upward.

Baudelaire's *flâneur* was, he claimed, the new observer and philosopher of modernity: "Sometimes he is a poet; more often he comes closer to the novelist or the moralist; he is the painter of the passing moment and of all the suggestions of eternity that it contains."[9]

The word *flâneur* is a noun derived from the verb *flâner*, to stroll. The *flâneur* in the new modern city is the person who observes humanity as he strolls through the city, not aimlessly, but with no preconceived plan. His walks take him wherever there are matters that interest him, as he studies humanity and learns about human nature. He is endlessly curious. He is observing and reflect-

---

9 Mayne, op. cit., 5.

ing on himself observing at the same moment. He is both present and an outsider. In short, he is someone who has a consciousness that is very modern.

Sometimes the *flâneur* is, as Baudelaire says, a poet or an artist. He mentions Daumier and Gavarni in this category. However, sometimes the *flâneur* is us—an observer of humanity, someone who is curious to learn about the human condition. Especially in the nineteenth century, without all of the technological distractions of our time, the *flâneur* will find in the street a kind of entertainment.

Baudelaire also revelled in the crowd, as Walter Benjamin emphasizes in his study of the poet. Crowds are a new phenomenon in the city. If you are in Paris, you must learn how to negotiate the crowd and the new boulevards. It is in the crowd that one can be incognito, anonymous, that one can observe: "For the perfect *flâneur*, for the passionate spectator, it is an immense joy to set up house in the heart of the multitude, amid the ebb and flow of movement, in the midst of the fugitive and the infinite."[10]

In the city there is not only modernity, there is a variety of human experience and lives found nowhere else. Baudelaire quotes from a conversation he had with Constantin Guys: according to Guys, anyone "who can yet be *bored in the heart of the multitude*, is a blockhead."[11]

With the reconstruction of Paris and the growth of the bourgeoisie, there grew public places where one could easily watch the human spectacle—cafes, theaters, and public parks, as well as in the streets strolling around. Sadly, given the social circumstances of the times and of later European history, it will not be until the last half of the twentieth century that we will have *flâneuses*.

What occurred alongside Baudelaire, who was a major art critic before he became a prominent poet, is that a number of artists in the next generation followed his lead in recording the Parisian lifestyle. They adopted Eugène Boudin's observation in 1868 about what to paint:

> The peasants have their painters, Millet, Jaque, Breton; and that is a good thing.... Well and good: but between you and me, the bourgeois walking along the jetty towards the sunset, has just as much right to be caught on canvas, 'to be brought to the light'... They too are often resting after a day's hard work, these people who come from their offices and from behind their desks... There's a serious and irrefutable argument.[12]

---

10 Mayne, op. cit., 9.
11 His emphasis, Mayne, op. cit., 10.
12 Eugène Boudin quote from Bonnard's letter to Ferdinand Martin, September 3, 1868; as cited in *Eugène Boudin*, G. Jean-Aubry with Robert Schmit, trans. Caroline Tisdall (Greenwich, Conn.: New York Graphic Society, 1968), 72.

The excellent impressionist painter Gustave Caillebotte loved Paris and painted its places and streets. In his *Young Man at His Window* (1875), Caillebotte also shows us how the new apartment buildings in the rebuilt Paris also functioned to permit the *flâneur* to stay at home and use the street as his own theater.

Fig. 1: Gustave Caillebotte, *Young Man at His Window*, 1876

Caillebotte painted his younger brother René in the family home in their apartment on the Rue de Miromesnil in the 8th arrondissement of Paris. Here, René looks out on the street to watch the parade of humanity pass by. Indeed, the chair behind him is facing the street, indicating that this window is not used for reading, but for perceiving and reflecting.

Caillebotte also introduces some sensitivities close to Baudelaire. René's back is to us, but his image is also reflected in the glass door to his right, indicating not only a realism attributed to Caillebotte by many critics, but a sense that his inner life is also part of this experience that can only be had in the modern city. René is alone in the midst of a million people and there is an aura of longing in the image as well as a sense of alienation.

A more public image of the *flâneur* is in Caillebotte's *Le Pont de l'Europe* (1876), the bridge in Paris over the Gare St. Lazare, the latter painted often by Monet and once, in a profound image, by Manet.

Fig. 2: Gustave Caillebotte, *Le Pont de l'Europe*, 1876

Here, the *flâneur* is the man in a top-hat, well-dressed, and he is accompanied by a woman. He is more interested in the urban environment than he is in the woman. Some commentators regard the man as Caillebotte and the woman as his companion, Anne-Marie Hagen. Others suggest the male is an anonymous *flâneur* accompanied by a prostitute, given that the location is near the train station. A third interpretation has the man as Caillebotte and the woman as someone he overtakes as he is walking; he turns to get a look at her.

The well-dressed male is not only a *flâneur*, he is also a dandy, another modern character admired by Baudelaire, who as a young man found himself in debt partly as a result of his profligate spending on fashion.

There is another observer in the image, a laborer gazing down at the Gare, indicating that one needn't be wealthy to engage in looking at the human theater. The street belongs to all in Paris. It is a most democratic place.

Two images from the theater at the same time tell us that one appeared in public not only to be entertained, but also to see and be seen.

Renoir's *La Loge* (1874) is delightful, and the woman is very beautiful. Her male companion uses his binoculars to look around, to see who is present and what they are wearing, and as well to perhaps eventually look at the stage.

Fig. 3: Pierre Auguste Renoir, *La Loge,* 1874

Mary Cassatt's *Woman in Black at the Opera* (1878)[13] shows a woman attending a matinee, perhaps alone, being one of the few in the image to be looking at the stage. She has not come to be admired but to experience an operatic production. Others look at others, and one male is stupidly leaning over the balcony, the better to view the enigmatic woman.

While writing the 50 prose poems collected in *Le Spleen de Paris*, Baudelaire also wrote more poems for the second edition of *Les Fleurs du Mal*, which would be published in 1861. Two of the poems for the second edition, *Le Voyage* and *Le Cygne (The Swan)*, both written in 1859, are reflections of Baudelaire's sensibility in modernity.

Voyaging is a theme often discussed by Baudelaire. In this poem, the voyagers set sail, not quite knowing where they will land, searching for something related to the infinite. Some are seeking, others are fleeing. The escape is from "*une patrie infâme*," a condition of civilization they detest—perhaps a desire to leave contemporary life—or from one's own past, or from a love affair or a relationship. They experience a variety of climates and conditions.

However, says Baudelaire, "But the true voyagers are those who move/simply to move—like lost balloons..." They are motivated by their own inner need, a quest for something which gives life meaning in the midst of the hardships of life. They must voyage. It becomes a way of living.

They seek El Dorados, but find reefs. They look for the exotic, but find the ordinary, even in the Indies and the Americas. A voice asks the travellers: "Tell us, what have you seen?"

They describe the beauty of the stars and the sea, and a variety of experiences gathered from the "chance countries" they have visited. The voice asks for more: "Yes, and what else?" And there is a sad response:

> Oh, trivial, childish minds!
> You've missed the most important things that we
> were forced to learn against our will. We've been
> from top to bottom of the ladder, and see
> only the pageant of immortal sin

The world is both different and the same when you travel. For all life on earth is filled with tyrants and selfishness, state officials with "dreams of power," workers who accept their "brutalizing lash," and religions "like our own/ all storming heaven." They have seen that which caused them to leave home in the first place. As in the whole of *Les Fleurs du Mal*, life is often harsh and melancholy.

---

[13] https://collections.mfa.org/objects/31365

Towards the end of the poem, Baudelaire makes reference to an old medieval legend, that of The Wild Hunt.

What now?, the poet asks:

> Shall we move or rest? Rest, if you can rest;
> Move if you must. One runs, but others drop
> And trick their vigilant antagonist.
> Time is a runner who can never stop

In the legend, the souls of the dead are moving all the time, very quickly, to nowhere. Anyone, totally exhausted, who stops running and leaves the crowd, crumbles immediately to dust. Baudelaire is here critiquing the problem of progress in modernity.[14] How do we deal with the need to always be in a state of Becoming? How do we handle modern consciousness? How do we cope with trying to find meaning in the contemporary world?

Baudelaire has no definitive answer to all his dilemmas. But he does have a position. Keep voyaging. Near the end the narrator in the poem says: "It's time. Old Captain, lift anchor, sink!/ The land rots, we shall sail…" The voyage is, of course, a metaphor for the questing life, even while knowing that we are to die. At the end: "Underworld or Paradise, no matter!/Through the unknown to find the *new*." [AH]

It is that nineteenth century logic, a dialectic, not one like Hegel or Marx, where it moves forward as if propelled by something outside itself. Rather, we decide on the voyage and we decide on the quest. Our experience is filled with paradox and contradiction. We have anxiety, but we keep moving. It is closer to later philosophy, to Camus' Sisyphus rather than to Condorcet's progress. Like Kierkegaard, Dostoyevsky and Nietzsche, Baudelaire will be claimed by the existentialists, though he wrote long before there was anything called existentialism.

One commentator notes that "*Le Voyage* is rightly regarded as one of Baudelaire's supreme achievements."[15] However, if there is a single poem that critics have agreed upon to be Baudelaire's "signature poem" it is the companion *Le Cygne* (*The Swan*).

The poem, in two parts which mirror one another, has about it an atmosphere of melancholy and grief, as Baudelaire tells us. As well, there is a sense of loss and a dialogue between the modern, memory, and the past.

---

14 Northrup Frye, *The Modern Century* (Toronto: Oxford University Press, 1990), 22–23.
15 Martin Turnell, *Baudelaire: A Study of His Poetry* (New Directions, 1954), 88.

It opens abruptly: "Andromache, I think of you." The poet asks us to think along with him, using the story of Andromache, the widow of Hector, to evoke mourning and exile, as she weeps beside a river. The narrator is in Paris, crossing the new Carrousel bridge, opened in 1834, and he tells us that memory is reminding him of what was lost. The old Paris is gone: "the face of a town/ Is more changeable than the heart of mortal man." He remembers what he used to see and laments. Now they are ghosts—a royal barracks, columns, a menagerie.

Then, suddenly, a new image. A swan escapes its cage, "web-footed on the dry sidewalk," and seeks water, stretching its neck to the sky, seeming to ask for rain.

The poet evokes Ovid and we now know what is happening. Andromache and the swan are partners in exile, as was Ovid, who in his last years was exiled by Augustus from Rome to Tomis, on the Black Sea, where he wrote his *Tristia*, a work full of loneliness and isolation. Baudelaire likens the swan searching the heavens for water to one or two of Ovid's tales in the *Metamorphoses*. The exile is tragic in the manner of Pheathon and/or Icarus, "trapped in a ruinous myth."

The second half of the poem also opens with a sharp image: "Paris change!" Here Baudelaire is dealing with one of the major issues in his poetry and prose. How do we cope with eternal change? Paris is "foundations, scaffoldings, tackle and blocks," all destruction and construction, all in its own metamorphosis.

By now the short poem is as dense as can be, filled with memory, loss, exile, grief, history, and allegory. The poet then inverts the order of the opening half. The swan comes to mind, then Andromache, and then of a "consumptive negress," herself in exile, seeking "The vanished coconuts of hidden Africa/ Behind the thickening granite of the mist."

What is lost cannot be regained, whether the Rome of its early days or the Paris of the days before Haussmann and Napoleon III. The sadness persists for the thoughtful poet: "I think of sailors washed up on uncharted islands,/ Of prisoners, the conquered, and more, so many more." We carry melancholy, loss, even tragedy, as we move through the new modern world.

The poem is dedicated to Victor Hugo, with whom Baudelaire had a somewhat turbulent friendship. Hugo was himself in exile, deciding after Napoleon III's coup at the end of 1851 to leave France in protest. In 1859 he was living on the nearby island of Guernsey, despite the fact that Napoleon III proclaimed a general amnesty in 1859 for political exiles, where he remained until the fall of Napoleon in 1870 as a result of the Franco-Prussian War.

In the letter to Hugo in which he sent him the poem, Baudelaire wrote:

> Herewith some lines written for you and with you in mind... I'll touch up the imperfections later on. What mattered to me was to express quickly all the suggestions sparked off by an accident, an image, and to show how the sight of a suffering animal pushes our thoughts toward all those who love, who are absent, who suffer, toward all those deprived of something they will never find again.

Hugo admired Baudelaire's *Les Fleurs du Mal* and several new poems that Baudelaire sent to him after the first edition of 1857, which were included in the 1861 edition. In a letter from Hugo to Baudelaire of October 6, 1859, Hugo famously wrote: *"Que faites-vous? Vous marchez. Vous allez en avant. Vous dotez le ciel de l'art d'on ne sait quel rayon macabre. Vous créez un frisson nouveau"* ("What are you doing? You proceed. You go forward. You endow the firmament with the art of a macabre beam. You create a new thrill" [AH]).

"*Un frisson nouveau*" became the shorthand for the evocation of the newness and uniqueness of Baudelaire's poetry. Yet, he went further in his experimentation. He decided that perhaps the new modernity required a new kind of poetry. Not only was the content of life new, but the way we write about it required a new type of literature, even different from the radical style of *Les Fleurs du Mal*. These came to be the prose poems of *Le Spleen de Paris*.

The hero and narrator of the prose poems is a *flâneur*, one who embodies both the voyage and the swan. However, the voyage is now in the contemporary city and the exile is someone who is distant, sometimes alienated, sometimes alone, sometimes heightened by his experiences. It is an urban voyage by someone who is an inner exile.

In 1862, 20 of the poems were published in *La Presse*, with a preface, which is in the form of a letter to the editor, Arsène Houssaye. Baudelaire notes that the order of the poems has "neither head nor tail," that is, each poem can stand by itself, and the order is arbitrary.

He then reflects on the form of the poems:

> Who among us has not dreamed, in his ambitious days, of the miracle of poetic prose, musical without rhythm or rhyme, supple enough to be adapted to the soul's lyrical movements, to the undulations of reverie, to the twists and turns that consciousness takes?

Baudelaire is here telling us he is moving into new territory, not in content but in form. How do we describe modernity? Perhaps we ought to do so in a "poetic prose," perhaps formal poetry is not quite adequate to the times.

It is not as odd as it seems. From 1857 Baudelaire and Flaubert, friends, were tied together in the public imagination because of the two trials. Flaubert is famous for the care in which he shaped his sentences and the rhythm of his language in *Madame Bovary*. To this day, his work is regarded as among the finest of examples of lyrical prose in any language. Moreover, we should remember that some prose forms were still relatively new in the mid-nineteenth century. The modern essay was invented in the sixteenth century, and the novel itself dates back to Cervantes in the early seventeenth century. The possibilities of prose were (and are) still being explored. Prose, of a certain kind, Baudelaire is claiming, may be the appropriate form for our times.

In the preface Baudelaire also tells the reader that his poems come from "my exploration of huge cities." He sought a "lyric prose" in order to convey his perceptions, reflections, and ideas. He knew he was doing something very different.

"Crowds" is one of the most revealing poems in relation to the voyage of the *flâneur*. "Enjoying the crowd is an art," he tells us. This new urban experience offers a variety of pleasures. In the multitude there is also solitude, they are not opposites. Indeed, the contradiction is a new kind of reality. Reflecting in the crowd you can enter into the lives of others: "The solitary, pensive walker finds a singular intoxication in this universal communion." This "singular intoxication" (*singulière ivresse*) is a "feverish pleasure," one which cannot be felt or understood by anyone locked up in themselves.

An element of chance is also part of the experience. You don't quite know when you will be lifted, and part of the joy is to encounter the unexpected. Welcome it, he asks of us. Moreover, he challenges the wealthy, who think that their good fortune means they are happy, and have what he terms a "stupid pride." Rather it is the voyagers who experience the greatest happiness. And he again uses the term intoxication, but with a different adjective. Yes, they are singular, but they are also "mysterious." Revel in it.

In another poem, "Windows," Baudelaire tells us of the interest in seeing what is behind a closed window as one strolls in the city. "There is no object more profound, more mysterious [again, mysterious], more fecund, more shadowy, more dazzling than a window lit by a candle." After all, who among us today has not been fascinated by a lit window at night, giving us a glimpse into the lives of others?

In the glimpse, the *flâneur* imagines a life by observing a gesture, a face, clothes, and other signals. This is important, he tells us, for this enables us to live and suffer with other people, to empathize with the human condition. Is it true? He asks. No matter, for it helps the voyager to feel alive.

An image used by both Baudelaire and Marx can be used to understand their different perspectives, even when there is a similar moral content to what they are doing. It is the halo.

In their *Manifesto of the Communist Party* (1848) Marx and Engels state in the section discussing the new bourgeoisie:

> The bourgeoisie has stripped of its halo every occupation hitherto honoured and looked up to with reverent awe. It has converted the physician, the lawyer, the priest, the poet, the man of science into its paid wage-labourers.[16]

For Marx, the halo gets in way of reality and is a function of class and power. He is happy that it has been ended in the new capitalist world and that the commodification of everything is becoming more evident. This, too, is part of the road to socialism.

In "Loss of a Halo," Baudelaire looks to experience, not economics. Two friends meet outside a bordello. One asks the other, who is a poet, what are you doing in this sort of place? You deal in "quintessences" and "ambrosia."

The poet tells the friend that something unusual occurred. He was in the chaos of the city, trying to cross a boulevard (a new boulevard of Baron Haussmann in Paris) amid all the dangers of encountering horses and carriages. He made a sudden movement, and his halo fell into the mud of the street, letting us know that in the modern city, halos and mud are often in the same place.

His response is different from the socio-economic one of Marx. I decided, says the poet, that losing my halo was a good thing. "Now I can walk about incognito, do vile things, and give myself up to debauchery, like simple mortals." I can be anonymous and thereby have the greatest variety of human experiences.

Get your halo back, says the friend. Get the police to assist you or put up an advertisement. The poet replies that his friend doesn't get it. I really like my new condition, he says. "Dignity bores me," for then I have to adopt a persona rather than being my own person. Besides, he says, I imagine that some awful poet will find it "and shamelessly put it on." Only fools quest for haloes.

Marx and Baudelaire both understand some of the contradictions and immorality of the bourgeois life. However, the former is interested in history, society, and political transformation. The latter explores experience and feelings. Both, it should be noted, care about the poor and the oppressed. Marx wants their status to change; Baudelaire wants to know about their soul. Each, in his own manner, helps us to understand the human condition.

In "The Soup and the Clouds" Baudelaire does acknowledge some of the realism of Marx, using an example that he had discussed in an essay several years earlier, *The Universal Exposition of 1855*. The prose poem is only two paragraphs. A man is looking at the sky while his beloved is preparing dinner and contemplates its great beauty, "almost as beautiful as the eyes of my beautiful beloved, my darling monstrous little green-eyed maniac."

---

16 Tucker, op. cit., 476.

He feels a punch in his back and his beloved, in a "charming voice... hoarsened by brandy," tells him: "So when are you going to eat your soup, you son of a bitch of a cloud merchant."

This is typical Baudelaire. Just when you think you have managed to pigeonhole him, he slips away, he makes you reflect again. The romanticism of the first paragraph gives way to the realism of the second one. We have to eat. You can't survive with your head always in the clouds.

This reminds us that, though Baudelaire saw himself in the Romantic tradition, he had admiration for Balzac, who he called a "great genius," and other Realists. In the 1855 essay he relates a story told about Balzac. The Realist writer found himself contemplating a beautiful painting, one of a winter scene in the countryside, including cottages and peasants, with smoke rising from the chimney of a small house. "How beautiful it is!" (Balzac) cried," reports Baudelaire. And then Balzac went on to ask: "But what are they doing in that cottage? What are their thoughts? What are their sorrows? Has it been a good harvest? *No doubt they have bills to pay?*"[17] The romanticism of a Constable painting will only get you so far. Sometimes, for Baudelaire as well, you need to think like Balzac, perhaps even Marx.

There is also a sadness in Baudelaire, sometimes redeemed by the night and by art. In "At One in the Morning" our *flâneur* desires nothing more than solitude. The first two words of the poem are, "Finally, alone!" The narrator is tired of humanity and has a double lock on his door to protect himself when he is at home. He longs to be separated from the rest of the world.

"Horrible life! Horrible city!" opens the third of four paragraphs. He goes on to lament that the experiences of the city sometimes force you into artificial "respectable" and foolish relationships and encounters. His life by day is fragmented and ordinary.

He wants to transcend the banality, "in the solitude of the night." How to get it? By producing "a few beautiful lines," through art, which will enable him to feel good about himself. Like the poet Ovid, whose work Baudelaire knew well, Baudelaire believes that art can transcend the banality of life. Perhaps, said Ovid, if I write great art, it can outlive the mightiest of empires. He did and it did.

There are a number of other important themes in *Le Spleen de Paris*. One of them is that of the role of chance or contingency in life in the new city. In "Crowds" Baudelaire talks about the joy of encountering the unexpected, the pleasure "in the unknown that turns up." Similarly, in "Plans" the *flâneur* has

---

17 *Baudelaire: Selected Writings on Art and Artists*, 120.

the realization that one needn't travel far, "why force my body to change its place, when my soul voyages with such agility." Just strolling in the teeming city gives him the opportunity, by chance, to find wonder and delight. He goes further in "Plans" by talking about plans as affording their own pleasure, a joy in anticipation.

There is another kind of chance mentioned in the well-known "The Eyes of the Poor." It is a poem that tells of the disappointment of a man in his lover because, while they are sitting in an elegant café, a poor family is outside gazing through the window at a world beyond their experience. "Get them out of here," says the lover, they disturb me. The man, who is reflecting on what happened, indicates in his story that her eyes, her essence, and by extension his as well, are "inhabited by Caprice and inspired by the Moon," that is, he knows and she does not that it is chance that put them inside the café and the poor family on the street. His empathy with the poor is one seen by Baudelaire as a recognition that life itself is contingent and arbitrary. This insight is so modern that it is one which is often used by the master of the short story of the twentieth century, Jorge Luis Borges. Baudelaire's view is one which can be found elaborated in Borges' "The Garden of Forking Paths" and other works.

The bizarre, the strange, even madness, all attract the *flâneur*. The normal is boring. Another famous poem, "Mademoiselle Bistouri,"[18] tells of him walking at the edge of town when a woman comes to him, puts her arm into his and says "you are a doctor, monsieur?" He says he is not and asks to be let go. Oh, no, she replies, you are certainly a doctor. Come to my home. The stroller is drawn to this possibility because "I dearly love a mystery." He permits himself to go to find what will happen.

It is a poor lodging (*taudis*, a slum), though there are hung portraits of famous doctors. And, he says "How pampered I was!" There is a good fire, wine, cigars. The woman keeps insisting he is a doctor and she endows him with a professional history as part of her own experience. She has a bundle of portraits of doctors and describes events in their lives. She builds a happy alternate world.

He asks her to tell when this obsession began. He tells us he had to persist in order to be understood, and he breaks the spell. "I don't know," she replies. "I don't remember."

---

[18] A bistouri is a surgical instrument, a knife whose blade is fixed or collapsible, and is used to make incisions.

He muses: "What bizarre things can be found in a large city, when one knows how to walk around and look for them! Life swarms with innocent monsters." He asks us to "have pity" on the mad and the words "have pity" are repeated for emphasis.

The poem usually placed after "Mademoiselle Bistouri," "Any Where Out of the World,"[19] is also about a kind of madness. The metaphor beginning the poem is one familiar in modern Western literature. "This life is a hospital," says the narrator. And we are patients "obsessed with switching beds."

The narrator has a conversation with what he describes as his soul, with himself. He asks whether he would prefer to go to other places—Lisbon, Holland, Batavia—but the soul has no response. He is numbed by his "disease," a disease of the soul. He thinks of other destinations, even as far as the Northern Lights. Finally his soul responds. Take me anywhere, "as long as it is out of this world!"

The mad move out of the world, as does Mademoiselle Bistouri. We have a need to escape, to find a place of solace, even if it is in an illusion or a dream. This theme is repeated often in *Le Spleen de Paris*. One poem, "Get Yourself Drunk," tells the reader to relieve the burdens of time, meaning that time and death is always the victor, the end of life. When you awake in the morning, he suggests, and look at your clock to find the time, have the will to have the clock answer: "It's time to get drunk." On what? And he repeats an earlier phrase, again for emphasis, "On wine, on poetry, or on virtue, whatever you like." Have passion, he is advising, have experiences which are sublime. That's the way to defeat the clock. As he says in the title of another poem, "To Each His Chimera," we need passion, even in illusion.

Baudelaire's love of crowds is matched by his interest and caring for the marginal in the new society of modernity. The poor get a great deal of attention in his prose poems. His reflections on the poor are filled with contrasts in condition and class, and the poor are given sympathy and dignity. One poem contrasts a wealthy child, beautiful, in lovely clothes, someone "in the absence of worry," in a park, with a lovely toy on the grass next to him.

"On the other side of the gate," as Baudelaire puts it, a place for others who are not of the bourgeoisie, there is a poor child, in a place with nettles and thistles, dressed poorly, dirty and puny. The rich child ignored his own toy, preferring to look at that of the poor child. That toy, in a box, is a live rat. As Baudelaire

---

[19] The title is in English, taken from "The Bridge of Sighs," a poem by Thomas Hood. In despair, a woman whose lover deceived her and whose family abandons her commits suicide by jumping into a river. She wants the water to take her "anywhere out of the world."

says, "The parents, through economy no doubt, had taken the toy from life itself." More significantly, the two children laughed together, as equals.

In "The Eyes of the Poor" there is a contrast between the muddy and dirty streets of Haussmann's renovating Paris and the new cafés and boulevards. The poor, staring from outside into the café, are dressed in rags. The inside of the café is decorated with gold cornices and lovely moldings, with images on the walls celebrating gluttony. The male narrator even remarks that as the poor are staring he felt ashamed of his glass, much bigger than his thirst.

In this case, Baudelaire alludes to two other new features of modernity. First, the lovers are not in some wood, alone, but conduct their relationship in public, on the streets, in the parks, at the theater, in cafés. Also, the relationship between the lovers has something political about it, something not part of traditional romance. Because the woman is so nasty to the poor, because she wants them "out of here," his view of her changes, as angelic and lovely as she appears to be. The first line of the poem is, *"Ah! vous voulez savoir pourquoi je vous hais aujourd'hui."* (Oh! So you wish to know why I hate you today. [AH]) *Je t'aime* becomes *je vous hais*.

In other poems, the narrator wants to lift the poor out of their despair. "Let's Beat Up the Poor" has a wonderful turnabout. A man hears an inner voice that suggests he do something to help the soul of a beggar who has "one of those unforgettable gazes that could overturn thrones," and who seems totally broken. He attacks the man, punches him. And then, what he calls a miracle does happen. The beggar turns on the man and attacks him, knocks out some teeth, blackens his eyes, and beats him with the branch of a tree. Wonderful, he says. Now the beggar had back his pride. So, he says, I told him: "Monsieur, *you are my equal.*" Honor me and join me for a drink.

Elderly women, often widows, are viewed with sympathy. He describes some of them in one poem as beggars who go outside tavern doors to gather up what crusts of bread might be there. He calls them "sixty-year-old gleaners," these city women, recalling for the reader the recent painting by Millet, *The Gleaners* (1857), in which Millet has sympathy for the women in the countryside who go to the fields to get what sustenance they can find after a harvest. The poet also has sympathy for a good old woman who wants to love a child who draws back in fear because of her decrepitude. In "Widows," he follows several women who try to find meaning in a life "with no friend, no conversation, no joy, no confidante."

The last woman he follows shows us that even Baudelaire can sometimes get it wrong. She is clearly poor, but has a kind of nobility in how she carries herself, very much in contrast with her poor surroundings. Virtue seems to be part of her person. She watched the world with interest. He wonders how she can be so different from so many others in her condition. Then he understood. The woman was holding the hand of a child. Still, the *flâneur* laments that the child will not give her the intimacy one needs. Rather, a child will be demanding, he says, selfish, and cannot be a confidante. Hence she too is alone.

He is wrong. For the child, who will grow up, gives his grandmother purpose, and she knows that she can help shape a life. That, my dear *flâneur*, is why she still has nobility.

Lovers get a lot of attention. Especially the eyes of lovers, which reflect their inner souls. Love is likened to a "natural opium," to dreams and utopia. "Invitation to the Voyage," is a poem dedicated to his lover, to how much love enriches life.

There is one poem, "The Generous Gambler," which has a Faustian edge to it. On a crowded boulevard the *flâneur* is jostled by *"un Etre mystérieux,"* a mysterious being who, though he had never seen him before, he recognizes. The being, the devil, asks him to follow and leads him to a luxurious underground dwelling, far more so than any Parisian home. The atmosphere was exquisite, beyond compare, so much so that one forgot all of the difficulties of ordinary life. The people underground had faces marked by a "fatal beauty," with a strange expression, including eyes that seemed dead and yet conveyed the horror of boredom and a longing to feel alive.

The narrator and devil become fast friends. They ate and drank wonderful wine and engaged in gambling. Of course, the *flâneur* lost his soul in the game, something that at the moment didn't bother him because he was drunk and his soul seemed often a useless and very annoying thing. The two converse, among other matters about different philosophies, and the devil betters any other conversationalist he had known. Indeed, the devil claims he wants the destruction of superstition. The only time he feared for his power, said the devil, was the day he heard a preacher, more profound than others, say about the Enlightenment that "the finest of all the devil's tricks was persuading you that he doesn't exist."

The narrator asks about God and if the devil had seen him recently. He answers indifferently. We acknowledge each other with a bow when we meet, he says. We are civil. However, we continue to retain the memory of old grudges.

They talk through the night. As the first light of dawn appears, the devil tells his friend that he wants him to think well of him, to know that sometimes he can be a *bon diable*, a good devil. Hence, he will provide him with compensation for the loss of his soul, in giving him what he would have won if chance had been on his side. Boredom, he says, "which is the source of both all your ills and all your miserable progress," will never again be a part of your life. All your desires will be fulfilled. And he leaves.

The *flâneur* tells us that at first he wanted to throw himself at the feet of the generous devil. However, he has second thoughts. Doubt appears, for he could not believe such happiness was possible. He tells us he went to bed, saying a prayer, "My God! Oh, my Lord, my God! Make the devil keep his promises to me."

It is ennui that is Baudelaire's greatest fear. This is a poet who not only helps us to understand modernity. He sought to find a way to transcend the despair that is part of the modern condition. He travels, he takes pleasure in his fellow human beings and their beauty and foibles. Baudelaire sought to make something beautiful, to be an artist with words. He helped shape the agenda for numerous others who follow.

In late 1867, possibly early in 1868, certainly after the death of Baudelaire on August 31, 1867, Édouard Manet wrote to Charles Asselineau, a friend of Baudelaire who would write in 1869 the first biography of the writer, offering two of his images for an edition of the poet's works:

> My dear Asselineau, I believe you are working at this time on an edition of the works of Baudelaire? If you put a portrait at the beginning of Spleen de Paris, I have one of Baudelaire with a hat, in other words, as a *flâneur*, which might not look bad at the start of this book. I also have another, bareheaded, more impressive, which would go well in a book of poetry. I would very much like to be given the task, and, of course, in proposing myself, I would give you my plates.[20]

Manet and Baudelaire met in 1859 and they became close friends, even though Manet was 11 years younger. They shared a desire to be in the present and to use their writing and artistic talents to examine it and understand it.

A friend and biographer of Manet, Antonin Proust, wrote about their friendship:

> For Manet the eye played such an important role that Paris has never known a flâneur like him, nor one who wandered about so effectively... He went to the Tuileries gardens almost everyday from two to four o'clock to draw *plein-air* studies under the trees... Baudelaire regularly accompanied him.[21]

Manet's first major painting was and is regarded to be *The Absinthe Drinker*, done in 1859, when he was 27. It is the first of several absinthe paintings by French

---

[20] Françoise Cachin, Charles S. Moffat, and Juliet Wilson Bareau, *Manet, 1832–83* (New York: Metropolitan Museum of Art, 1983), 158.
[21] Patricia A. Ward, *Baudelaire and the Poetics of Modernity* (Nashville: Vanderbilt University Press, 2001), 45.

Fig. 4: Edouard Manet, *The Absinthe Drinker*, 1859

painters in the last half of the nineteenth century, a forerunner to the very famous study done by Degas in 1875–76.

The painting is large (about 71 inches high and 41 inches wide). It presents a real person, a Parisian *chiffonier* (rag-picker) named Collardet, who was to be found in the area around the Louvre. Collardet is given a full length portrait, much like other painters would give to their bourgeois subjects. He is also given dignity, much as Baudelaire gave dignity to his widows, children, and the poor. Instead of having his absinthe drinker seen to be a model of the

Fig. 5: Honoré Daumier, *The Exposition of 1859*. "My dear, since we won't have enough time to see everything in one day, you should look at the pictures on the right while I'll look at the ones on the left, and when we get home, we'll tell each other everything we saw." Baudelaire knew Daumier and admired his work. In 1865 the art critic Champfleury, who was preparing a work on caricatures, asked Baudelaire to write some verses as an appreciation of Daumier. Baudelaire responded quickly with *Verses for Honoré Daumier's Portrait*.

harm of the desolate life, Collardet is dressed as a dandy, with a top hat and a cloak. An empty bottle is at his feet, but he still stands more like a bourgeois than a sad drunk.

It is, for the later Manet who was an Impressionist, a dark image, painted in brown and grey tones, resembling some of the work of Gustave Courbet and the Realist school. Manet's most important influence was his deep admiration for the work of the Spanish master, Diego Velásquez. He said, "I painted a Parisian character whom I had studied in Paris, and I executed it with the technical simplicity I discovered in Velázquez."[22]

Manet submitted the painting to the Salon of 1859. It was rejected. The only juror voting in its favor was Eugene Delacroix, who, incidentally, was the subject of high praise by Baudelaire in his essay *The Life and Work of Eugène Delacroix* (1863). The jury described the painting as one of "vulgar realism."

The "vulgarity" was of course in not depicting Collardet as a hopeless, immoral, unredeemable human being. Manet had been warned. As the painting neared its completion, Manet showed it to his teacher, the famous Thomas Couture, who was repelled. Couture's opinion was that of the academic establishment. He called it an abomination: "My poor friend, you are the absinthe drinker. It is you who have lost your moral sense."

The result was a *succès de scandale* when it was shown, the first in a series of them as Manet continued his quest to tell about the present. As he said later, in 1865, in a letter to Baudelaire, "Insults are pouring down on me as thick as hail."[23]

The painting Manet referred to when suggesting Asselineau might use it for the cover of Baudelaire's works was his *Music in the Tuileries* (1862), a direct result of the habit of going to the Tuileries Gardens, sometimes accompanied by Baudelaire, to paint *plein-air*.

Again, it is very different from the academic painting of the time. Like Baudelaire's poems, it is fully in the present. Now, in the modern city, there are public parks and, on occasion, public concerts. The bourgeoisie would attend, probably on a Sunday, not necessarily to listen to the music, but to see and be seen, to greet friends and acquaintances, and to exchange news and gossip.

The focus of the painting is not on the concert, it is on the people. Like Courbet's earlier Realist painting, *Funeral at Ornans* (1849–50), which treated a funeral as a social event, Manet is interested in the gathering, the sociology of the place. It is a kind of group beloved by the *flâneur*, a crowd.

The painting is organized somewhat off center, with the seated women, children playing at their feet, dressed in gold, immediately capturing our attention. The tree to their left, which divides the canvas, is also not centered. No golden triangle here.

This group is fashionable. The men's clothes and hats are those of the dandy. The women's dresses are elegant. The children are well-dressed.

---

[22] Pierre Schneider, *The World of Manet* (New York: Time-Life Books, 1968), 22.
[23] *Manet by Himself*, ed. Juliet Wilson-Bareau (London: Macdonald and Co. Ltd., 1991), letter of March 25, 1865, 33.

Fig. 6: Edouard Manet, *Music in the Tuileries*, 1862

Manet made certain that the painting reflected not simply society, but the artistic world. Baudelaire is directly behind the blue hat of the seated woman on our left. Henri Fantin-Latour, the bearded man, is next to Baudelaire. Théophile Gautier is the third person in that group. Manet's brother Eugène is in profile in the center. Behind Eugène, seated next to a tree, is the prominent composer Jacques Offenbach. Manet painted himself into the image. He is on the far left, in profile, partly cut off. Next to him is Albert de Balleroy.

It is orderly and disorderly at the same time, like life. The metal chairs are strewn about. A parasol is lying unused, some of the crowd cannot be clearly seen. It is not posed at all.

In 1846, Baudelaire said, "Parisian life is rich in poetic and wonderful subjects. The marvelous envelops and saturates us like the atmosphere; but we fail to see it."[24] Manet saw it and in this image decided he had to move on to a new kind of painting in order to record it. The brush strokes are sometimes bold, sometimes hazy. The painting gets darker as you move further into the scene, even from left to right, giving us the perspective of the artist.

Manet knew well the history of art in the West. He also was determined to bring art into modernity. It was Baudelaire who, in 1855, said, "Like all my

---

[24] *The Salon of 1846*, 107 in Charvet, *Baudelaire: Selected Writings on Art*.

friends I have tried more than once to lock myself inside a system, so as to be able to pontificate as I liked. But a system is a kind of damnation that condemns us to perpetual backsliding."[25] As Dostoyevsky would remark in 1864, "I agree that two and two make four is an excellent thing; but to give everything its due, two and two make five is also a very fine thing."[26]

The system Manet inherited could not provide him with the tools to paint his experience. He began to invent a new way of painting in this image. It is, says one prominent critic, "the first truly modern picture, the first to embody fully what Baudelaire called the 'essential quality of being present.'"[27]

Fig. 7: Claude Monet, *Impression, Sunrise,* 1872

The term Impressionism, to describe the new style, was not invented until over a decade later, in 1874. In response to a viewing of Claude Monet's *Sunrise, Impression,* painted in 1872, part of an exhibition he insultingly called *The Exhibition of the Impressionists,* the critic Louis Leroy said of Monet's image, "Wallpaper in its

---

25 *The Universal Exhibition of 1855,* in Charvet, 117–118.
26 *Notes from Underground* (Penguin, 1972), 41.
27 Schneider, op. cit., 24.

embryonic state is more finished than that seascape." It can easily be argued that *Music in the Tuileries* is not only the first modern painting, it is also the first Impressionist painting.

Manet not only went on to record the present. Of all the Impressionists he is the one who most critiqued it. Many of his images are of happy people engaged in activities such as boating, on holidays, in cafes, and the street. However, some of his work also steps back and asks us to reflect on the limits of modernity itself. Three of his most famous works, *Luncheon on the Grass* (1862–63), *La Gare St. Lazare* (1873), and *A Bar at the Folies-Bergère*, deal profoundly with the status of women in his time.

Luncheon on the Grass (*Le déjeuner sur l'herbe*) was the greatest art scandal of the Second Empire. In it, Manet, as in some of his other works, got inspiration from the artistic tradition, then brought the subject into modernity. There is a picnic in a public place, again something that is part of modern life. However, the woman in the foreground lacks clothes, while her two male friends are dressed in the dandy style of the day. In the background a woman is bathing (Manet initially called the painting *The Bath*). As in *Music in the Tuileries*, there is a patch of sky at the top-center, which prevents the painting from being limited by the trees and branches. The canvas is very large, 6.8 feet by 8.7 feet, producing the effect of an epic or myth.

The two images Manet used as guides were part of the Western canon. One was Giorgione's *Fête Champêtre* (c. 1509), which provided the subject; the second was Raimondi's engraving of Raphael's The *Judgment of Paris* (c. 1510–20), which gave him, in its lower right, the arrangement of the three main figures.

Manet submitted *Luncheon* and two other images for entry into the Salon of 1863. They were refused by the judges, who refused entry to more than half of the 5,000 works submitted by various artists. It should be remembered that this was before there were art galleries and other outlets which were accepted by the public as places to view and purchase living artists. Some famous painters could reject using the Salon. But new artists had few alternatives. The Salon judges were the gatekeepers.

However, there were so many complaints about the judges' decisions that the authorities decided to permit rejected artists to show their works in an alternate exhibition, which became the *Salon des Refusés*. Among others, Courbet, Whistler, Pissarro, Cézanne, Fantin-Latour, Jongkind, and Manet showed their works in the *Refusés*, which opened just after the establishment salon did.

Manet's *Luncheon* was the scandal of the show, far more so than the earlier *Absinthe Drinker*. Many who went to view the collective show did so to disdain and insult these artists who they decided had little talent and nothing to offer. While a few praised *Luncheon*, most found it "immoral," "indecent," "shame-

Fig. 8: Edouard Manet, *Le déjeuner sur l'herbe*, 1863

less," "a joke," "an absurd composition." One critic remarked, "This art... isn't healthy."

What was the problem? Part of it can be attributed to the composition. In a manner similar to that of the *Music*, Manet's brush strokes were sometimes patently present, there seemed to be a juxtaposition of styles—a still life in the lower left, a woman painted as simply unclothed, the dandies relating to one another and not the woman, the bather seemingly in almost a second painting attached to the first.

But the bigger problem for the audience, which was the aristocrats and the bourgeoisie, was the woman herself. In Giorgione's image there are nudes, and they don't shock (to use Walter Benjamin's term). They are simply idealized women, representing the beauty of the human form, something common in

Western art from the Renaissance forward. However, Manet's woman[28] is a real person, not some Platonic ideal. And she is not nude, she is naked.

Further, the gaze of the woman is unusual for a portrait. She is looking directly at the viewer, neither startled nor ashamed, simply present at the time the viewer appears. In fact, her gaze tells the viewer that she knows he is there and is voyeuristically looking at something he decides is erotic, perhaps even pornographic. She is telling us that she will not be what the marketplace of opinion wants her to be. She is asserting her freedom to be what she chooses, regardless of the criticism. Like Manet's *Olympia*, which was shown in the Salon of 1865, *Luncheon* portrays a woman who, looked down upon by society, is bold and independent.

The status of women and the limitations they face are central to Manet's *La Gare St. Lazare* (sometimes called *The Railway*) of 1873.

Railways were viewed positively by the Impressionists and others, part of the power and new opportunities of modernity. Indeed, railway stations in the nineteenth century were citadels, constructed to celebrate the new modern age. One, the Gare d'Orsay in Paris, opened in 1900, has been converted into one of the great museums of nineteenth century art and sculpture, now the Musée d'Orsay.

The most famous set of French railway paintings are those of the Gare St. Lazare by Claude Monet. In 1877, between January and March, Monet did a dozen images of what was the oldest railway station in Paris, at the time serving over 13 million passengers a year. The area around the Gare St. Lazare was a favorite of the Impressionists, including Caillebotte.

Monet's images are about atmosphere. Smoke rises and limits what we can see, but it also suggests the mystery and wonder of travel to new and interesting places. There is the iron and the metal canopy, celebrating the industrial revolution and its possibilities. The apartments of the new Paris of Haussmann are in the background.

Manet's *Gare St. Lazare* takes a very different perspective.

Here, it is from the perspective of two females outside the Gare. The young girl is dressed well, indicating that she is of the bourgeois class. The elder, looking at us, is likely a governess, perhaps a parent.

They are behind the fence, symbolically behind bars. That is, they are barred from the experience of the station or travel because they are respectable females who, at that time, could not travel by themselves.

---

28 The model was Victorine-Louise Meurent, then Manet's regular model, who later became a painter good enough to be accepted into the *Société des Artistes Français* in 1903.

Fig. 9: Claude Monet, Le Pont de l'Europe – Gare Saint Lazare à Paris, 1877

The woman has her back to the fence, knowing that she cannot participate in the adventure of the railway. She wears a dark blue twill demure dress, but there are indications of some sensuality in her loose red hair and the black velvet band around her neck, like that of the sexual figure in Manet's earlier *Olympia* (1863). She has in her lap a small dog, a symbol of domesticity.

The woman also has an open book in her hands, a perceptive insight by Monet. He is telling us that females, at that time, can only travel in two ways. First, in their own minds, recalling the description in Flaubert's *Madame Bovary*, when Emma takes out a map of Paris and traces with her finger possible walks. As well, women can travel through books, as they do in many European novels of the time.

The girl is still innocent enough to try to look into the railway. However, for Manet, unlike Monet, the smoke is used as fog, as something that limits possibilities.

There is another interpretation to be considered. Some critics see the young girl and the mature woman as the same person. She is hopeful and optimistic as

Fig. 10: Edouard Manet, *Gare Saint-Lazare* (also called *Chemin de Fer* or *The Railway*), 1872

a youth, but she will become the domesticated governess when she is an adult as a result of the constraints of bourgeois society.

In 1977, the outstanding historian of the Renaissance, Joan Kelly-Gadol, asked a seemingly simple question that resonated throughout the discipline: "Did women have a Renaissance?" She concluded that events that further the development of men, that free them from traditional constraints, might have a different effect on women in some periods. In evaluating this for the Renaissance she concluded that "there was no renaissance for women—at least not during the Renaissance."[29] That conclusion, even the fact that the Renaissance affected women adversely, has come to be accepted.

Manet, in his images related to women, is implicitly asking a similar question: "Did women benefit from modernity as it was introduced in the nineteenth

---

29 R. Bridenthal and C. Koonz, *Becoming Visible: Women in European History* (Boston: Houghton Mifflin, 1977), 139, 161.

century?" His answer is clear. No. From the perspective of our time, we can echo Kelly-Gadol and say there was no modernity for women in Baudelaire's and Manet's time. Women arguably may have entered modernity, but not until at least a century later. We should note that other major 1859 figures said much the same thing, though most males simply didn't even think this was a matter to be considered. John Stuart Mill argued in many places for the equality for women, most notably in his *The Subjection of Women* (1869).

Flaubert's Emma Bovary tries to rebel and fails, but she becomes a sympathetic figure. In the novel Flaubert describes Emma's desire for the glamor and experience of Paris. She is stuck in a tiny town near Rouen, and longs for something more:

> What was this Paris like? What a vague name! She repeated it in a low voice, for the mere pleasure of it; it rang in her ears like a great cathedral bell; it shone before her eyes, even on the labels of her pomade-pots.
>
> At night, when the carriers passed under her windows in their carts singing the "Marjolaine," she awoke, and listened to the noise of the iron-bound wheels, which, as they gained the country road, was soon deadened by the soil. "They will be there to-morrow!" she said to herself...
>
> She bought a plan of Paris, and with the tip of her finger on the map she walked about the capital. She went up the boulevards, stopping at every turning, between the lines of the streets, in front of the white squares that represented the houses. At last she would close the lids of her weary eyes, and see in the darkness the gas jets flaring in the wind and the steps of carriages lowered with much noise before the peristyles of theatres.
>
> She took in "La Corbeille," a lady's journal, and the "Sylphe des Salons." She devoured, without skipping a word, all the accounts of first nights, races, and soirees, took interest in the debut of a singer, in the opening of a new shop. She knew the latest fashions, the addresses of the best tailors, the days of the Bois and the Opera. In Eugene Sue she studied descriptions of furniture; she read Balzac and George Sand, seeking in them imaginary satisfaction for her own desires...
>
> Paris, more vague than the ocean, glimmered before Emma's eyes in an atmosphere of vermilion... All her immediate surroundings, the wearisome country, the middle-class imbeciles, the mediocrity of existence, seemed to her exceptional, a peculiar chance that had caught hold of her, while beyond stretched, as far as eye could see, an immense land of joys and passions.[30]

Manet's last major work was the *Bar at the Folies-Bergère* (1882), one of his most famous paintings.

---

[30] https://www.gutenberg.org/files/2413/2413-h/2413-h.htm#link2HCH0009. The translation is by Eleanor Marx-Aveling, who was the daughter of Karl Marx.

Fig. 11: Edouard Manet, *Bar at the Folies-Bergère*, 1882

The Folies-Bergère was not a simple café. Rather, it offered glitter, luxury, and entertainment. Here we see the customers and the café through the device of the mirror, as the barmaid sees them. There is an acrobat in the upper left, the elegant chandelier, the lights, and the well-dressed patrons. On the marble bar there are bottles to drink, flowers, and oranges. The young barmaid is dressed well, undoubtedly a costume or uniform of the establishment.

Yet there are two enigmatic matters in the painting pointed out by most commentators. First is the fact that the barmaid has an expression described variously as tired, bored, sad, or unhappy. She is not joyful in this place of entertainment and, in the moment Manet paints her, she does not pretend to be joyful, as her position requires.

As well, there is a painterly matter. To the right we see the back of the barmaid who is facing a gentleman. However, that is not where the barmaid should be in the mirror and the gentleman is far larger than he should be if he is in the mirror.

What is happening here is that Manet, having earlier broken a number of the rules of academic painting in his brushstrokes, perspective, and in the presentation of a finished painting, here took his modernity one step further. There is not one image in the painting, but two. There is the barmaid, the mirror, and the luxuriousness of the *Folies-Bergère*. Then there is the encounter of the barmaid and the gentleman.

He pushed the painterly enterprise beyond the rule that you paint what you see and you paint a single image on a single canvas. This had been done before in the West, often on triptychs in the late Middle Ages and the Renaissance, where on a single panel there might be two or more pieces of a story. Now, he moves beyond appearance to consciousness.

The barmaid is indeed very sad at this moment because she is contemplating her fate later in the night, when she will be expected to entertain a gentleman as part of her employment. Some commentators tell us that she is a prostitute, but what choice does she have? She is poor, she is not well educated, she is not of the bourgeoisie. Manet paints her as a subject. To the various "gentlemen" she is an object.

This, too, explains her visage. This is, among many other things, Manet's equivalent of Baudelaire's "Eyes of the Poor." However, whereas Baudelaire's poor family marvel at the glitter, our barmaid is despondent, for she can never share in it, and she is alienated as a result of her circumstances.

Baudelaire liked using eyes as an indication of mood, feelings, and personality. The barmaid's eyes are not simply the eyes of the poor. They are the eyes of a woman who is poor, exploited, and abused. The naked woman in *Luncheon* looks directly at the viewer, as does the governess in *Gare St. Lazare*. They manage to have agency and dignity in the face of their difficult world. Our sad barmaid cannot manage that, but we as viewers can sympathize with her plight. Modernity leaves her out.

Baudelaire and Manet both broke new ground in their quest for a way to understand modernity and to record it. The two friends, one in poetry and the other in art, were revolutionaries in their fields. They changed the way poets and artists worked after them and gave them tools to record and understand this new world and its experiences. Neither poetry nor painting nor philosophy would be the same after these two friends, who went to the Tuileries together, and who died too young, provided us with their insights.

Chapter II
Giuseppe Mazzini: Prophet of Nationalism

"Your country should be your Temple," proclaimed Giuseppe Mazzini in *The Duties of Man*, published in early 1860 in the midst of the creation of modern Italy.

1859 was a key year in the history of nationalism, with the Italian wars inaugurating a decade when new nation-states came into being and others faced crises of existence (most notably the United States, an offshoot of Europe).

Modern nationalism is a creation of the late eighteenth and early nineteenth centuries. It was a reaction against the cosmopolitan and universal ideas of the French revolution, culminating in Napoleon attempting to turn Europe into a single empire. Some states, Russia and Spain for example, fiercely defended their national identity, not merely the legitimacy of the ruling dynastic family.

Nationalism had strong theoretical underpinnings in the work of German philosophers such as Herder and Fichte, who responded to the universal claims of the Enlightenment *philosophes* with a new idea. There are multiple expressions of human culture and society, they asserted, each providing benefit to the totality of human experience. Germans, Russians, and others should cultivate their own culture and modes of living, thereby asserting the validity of their own identities, instead of wanting to emulate Frenchmen. It was a new celebration of multiplicity and was reflected in the growth of Slavophile expressions in Russia, the pride in German romantic works, and the assertion by Mazzini and others that there was something called Italian identity which should be nurtured and valued.

The most revolutionary development of this idea came in the works of the most influential European philosopher of the early nineteenth century, G.W.F. Hegel (1770–1831). Hegel argued that the idea of freedom is only realized in its actualization in the state. It is he who put forth the idea, one of the most powerful in the modern world, that the state is not just some accidental growth in history, but that each nation needs a state to protect and further its nature and spirit:

> [The State] is the realization of Freedom, of the absolute, final purpose, and exists for its own sake. All the value man has, all spiritual reality, he has only through the state... Only thus is he truly a consciousness, only thus does he partake in morality, in the legal and moral life of the state... The state is the divine idea as it exists on earth.
>
> Thus the state is the definite object of world history proper. In it freedom achieves its objectivity and lives in the enjoyment of this objectivity...

This idea of the state did exist in the ancient world, in the history of the Jews and in the Greek idea of the polis. But it was replaced by several developments: universals—such as the Roman Empire and the Church Universal; localisms—identity and order found in local culture as the feudal system developed; or patriotism—loyalty to a dynasty or ruling house.

Mazzini himself believed that before 1859 only two European states, Britain and France, reflected the idea that the nation and the state should be one.

The idea of the nation carried with it another revolutionary notion. It argued that one's identity was now rooted in the nation to which one belonged. In an age when religious authority was waning, this notion became a powerful way of understanding the self and one's place in society and history. No longer were you simply a Christian and/or a cosmopolitan rational human being. Now you could argue you were a Russian, a Spaniard, an Italian, a Frenchman. Of course, it could become far more complicated than it seemed, for many in Spain would claim to be Catalan, Basque, Aragonese, etc., and the same held true for other areas. For example, for all of Mazzini's claim that there was something ideational called Italy, most on the Italian peninsula in 1859 related to their local culture, be it the Abruzzi, Naples or Lombardy. How to organize it remains to this day something to be argued about.

When the nation was called into existence as an important concept, three main criteria were cited. First, there must be a group bound together by language. Then, the group's identity must derive from a common history. And finally, the group must have a common culture.

Often tied to the three criteria was the belief that a certain piece of land was sacred to the nation, that some land had a mystical quality central to a people. Hence, when nationalists talked about their identities—be they Russians, Germans, the English or the French—they included a geography in addition to language, culture, and history. In the mid-nineteenth century, to be an autonomous nation meant to have a state that included land considered traditional, even holy, to its people.

In 1859, there were many nations in Europe, but there were not many nation-states. In the multi-national Austrian Empire, for example, neither Austria nor many other groups, including Hungarians, Poles, Romanians, and Czechs, were nation-states. Italy was, as Metternich had called it, "a geographical expression" and had many local cultures that could conceivably call themselves nations. There were 39 German entities after the Congress of Vienna, the two largest being Prussia and Austria. The Balkans, with Bulgarians, Croats, Albanians, Serbs, and others, were under the power of the Ottoman Empire. Greece and Russia might also be included, as well as Portugal and Spain. On the other hand, Mazzini did recognize Britain and France as nation-states.

Long before Benedict Anderson in 1983 styled nations as "imagined communities," it was an accepted understanding that the nation as identity had a kind of ontological reality. For those who came to believe that being part of a nation was important, it was the nation that provided identity and meaning. This shift in identity occurred alongside traditional religious notions and concepts of a universal human nature, and it came to be a powerful concept.

The idea of the nation-state also came to be tied to concepts about the legitimacy of states. Heretofore, what made a sovereign entity legitimate was tradition—as was agreed upon yet again at the last major peace conference which reorganized Europe, the Congress of Vienna of 1814–15, following the defeat of Napoleon.

Nationalism changed all such thinking. Now it was argued that legitimacy came from binding like people together in a state which reflected and forwarded their national identity. Mazzini argued that there is such a thing as Italian identity and that it is imperative, therefore, to have an Italian state. Similar arguments were made about Germany and other groups. Hence, not only did nationalism provide a new mode of identity, it also gave those who did not have a state a teleology, a sense of an appropriate future, as a people bound together in a single state.

Other developments aided this new idea. History, and national history, were now seen to be important—leading to the advancement of history as an academic discipline and to the beginnings of the teaching of national history as part of the ordinary curriculum.

Of course there was much fine and influential history written prior to the nineteenth century, from Herodotus to Voltaire. However, it was in Germany in the early nineteenth century that history became a *wissenschaft*, a systematic and organized study. At Berlin, the graduate history seminar was established by such people as Leopold Ranke, recognized as the founder of the discipline. Here method was taught, rules of evidence were codified, and professional historians were trained.

What was most important in the coincidence of the founding of history as an academic study and the rise of nationalism was that the discipline, from its early years, made politics and the state the center of its interest. What is to be studied? The growth of the state. How do you do this? By analyzing political and diplomatic documents.

As history is practiced today there are many branches and many units of study—some relate to gender, many to society, others to political economy, even some to feelings or food or the environment. Not so in the nineteenth century. For the vast majority of historians, the past was politics and diplomacy and the ontological reality were the state and/or the nation. Even today, when the discipline has become far broader, most academic departments of history divide the universe according to states or regions. And, of course, in virtually every state and/or nation, national history today is a required study for all in their educational systems.

History as significant in the nature of human beings and their sense of belonging—nationalism—and history as a universal course of study gave rise in the nineteenth century to what is called historicism and has since become part of modern consciousness.

Historicism posits that nothing can be understood without knowing its past—be it human beings, nations, geological formations, planets, or species. Hence, to be an Italian is to understand how Italy today came to be what it is; to be a member of a species is to realize that you came into being as part of a historical development.

If nationalism is modern, so is historicism as a necessary mode of understanding. Both share the major characteristic of modernity, that of seeing the world as in process, as Becoming rather than Being.

Historicism as a mode of analysis and part of epistemology received legitimacy and impetus from a number of the giants of nineteenth century European thought. Hegel introduced the dialectic and emphasized process. Marx is nowhere without history, without knowing where one is in the course of development. "Men make their own history," he famously said in 1852, "but they do not make it just as they please; they do not make it under circumstances chosen by themselves but under circumstances directly found, given and transmitted from the past."

Another figure of 1859, Charles Darwin, transformed our understanding of human nature and nature itself with his theory of evolution, an idea which takes historical development as central to its methods of analysis and understanding. We are who we are as a result of what we have been.

Other system makers of the century were historicists, including August Comte, Herbert Spencer, and Giambattista Vico, that eighteenth century figure who disappeared from the intellectual discourse before being resurrected in the first half of the nineteenth century. Looking forward with regard to human behavior, Freud founded his personality theory on historicist premises. We as individuals are who we are in large part because of our past.

At the outset of 1859, what we call Italy was a peninsula divided into eight different entities. In the north there was the independent sovereign kingdom of Piedmont-Sardinia, and two areas, Lombardy and Venetia, which were part of the Austrian Empire. In the south there was the large Kingdom of the Two Sicilies, including the island of Sicily and the large area on the mainland whose center was the city of Naples. This latter entity was ruled by the Bourbons and had family and historical ties to Spain.

The center of the peninsula included several small entities—Parma, Modena, and Tuscany. The major entity in the center was The Papal States, a sovereign state with Rome at its head, which bisected the whole of the peninsula and which was ruled by the Papacy since the eighth century.

Italy had not been united since the fall of the Roman Empire. Clear divisions existed. The south, the *mezzogiorno*, was poor and agricultural. The north, while also agricultural, was better off and would later begin to industrialize. Culturally, it could easily be argued that there was very little that some areas had in common with others. For example, the language, history, and culture of Venezia was as different from that of Sicily as it was different from, say, Spain. In 1859 there was something artificial in claiming that the whole peninsula had a destiny to be a nation-state. Very few would have predicted in early 1859 that Italy, with a tiny exception, would be united as a sovereign entity by 1870.

Mazzini (1805–1872) made himself into the most prominent nationalist of his time. His mission of Italian independence and unity was conducted both in action and thought, as he worked tirelessly for his cause. In 1829 he joined the *carbonari*, a set of secret societies that had as their goal a greater autonomy for the Italian sovereignties under foreign rule, and found himself imprisoned in Piedmont, a very authoritarian government, in late 1830.

Released after three months, he became an exile for the remainder of his life, appearing in Italy only for short periods of time. He lived in Marseilles at first, and there, with a group of other exiles, founded Young Italy, an organization which has come to be regarded as the first Italian political party, whose goals were republicanism and nationalism.

While in Marseilles, Mazzini displayed those personal qualities which made him a force. He was handsome, even described by some as "beautiful," extraordinarily eloquent, highly intelligent, well-read, and charismatic. He quickly became the leader of the movement, which also meant that he was spied upon from this moment on by governments fearful of his work and his goals.

Young Italy began a journal with the same name, regularly smuggled into Italy, whose contents were mainly written by Mazzini. It served as both a platform and a way of recruiting others into the cause. In 1833 there was an attempted coup in Piedmont by military officers supporting some of the ideas of Young Italy. The rebellion failed—as a result a dozen people were executed in a public event, many were imprisoned, many went abroad to escape death or prison. Mazzini was one of those found guilty and was given a death sentence *in absentia*, not the last time he would avoid persecution by being abroad.

He was expelled from France and moved to Switzerland for the years 1833–1836. There he continued his underground activity, focussing on fomenting a revolution in Savoy, whereby the populace would liberate itself from the then authoritarian Piedmontese regime. Here, he consolidated his reputation as someone always dangerous to authority, as one who realized early that only through struggle could liberation be realized.

In Switzerland, in 1834, he was joined by other refugees from Italy, Poland and the Germanies in founding a pan-European group, Young Europe. The idea was to finish, in his mind, what had been started by the French Revolution of 1789. That event gave to Europe the idea of individual liberty; now, thought Mazzini and others, it was time for a second major revolution which would bring liberty to the nations of the continent. Here began Mazzini's idea that free nations would live in peace and through them create a loose federation of Europe itself to look after continental matters. Mazzini could lay claim to being among the earliest supporters of a European Union, along with the French utopian socialist Henri de Saint-Simon, who published his thoughts on this goal more than a decade earlier.

In 1836, the Swiss government followed the example of the French three years earlier, and declared Mazzini *persona non grata*. Some tried to protect him and permit him to stay, but they were unsuccessful. For six months he was an underground man, hiding from the authorities, living surreptitiously. Finally, in 1837, he decided to move to England, where he could live and work openly.

He resided in London for most of the rest of his life. In London, Mazzini became the most famous and courted exile of his time. John Stuart Mill showed him respect and he became close friends with Thomas and Jane Carlyle. He lived close to poverty, but this only gave cause to further celebrate his integrity and dignity. Invited to the homes of prominent Londoners, he was regarded as a person of importance. Carlyle wrote of him: "He had great talent, certainly the only acquaintance of mine of anything like equal intellect who ever became entangled in what seemed to me hopeless visions… He had fine tastes, particularly in music. But he gave himself up as a martyr and sacrifice to his aims for Italy."[31]

Metternich and others regarded Mazzini at this time as the most important revolutionary in all of Europe. He re-launched Young Italy in 1839 and was involved in a number of schemes opposing outside authority on the Italian peninsula.

---

31 Dennis Mack Smith, *Mazzini* (Yale University Press, 1994), 31.

Mazzini's most prominent involvement came as a result of the uprisings of 1848. In Rome, Pope Pius IX fled the city, and in 1849 Mazzini led the short-lived Roman Republic in a government that was regarded as capable and efficient. The French helped to restore Pius IX to his throne and the pope, thought to have liberal tendencies when he was elevated to the papacy in 1846, became the Roman pontiff most opposed to all of modernity until his death in 1878 ended the longest reign in papal history.

Mazzini's reputation was only furthered by his efforts in 1848–49, but again he went into exile and organized a series of conspiracies in favour of Italian independence from abroad. He was the touchstone for many conspirators and actors, and it is agreed that his contribution to Italian unity and independence was necessary for it to happen.

Necessary, but not sufficient. In 1858 and 1859 Cavour in Piedmont and Garibaldi from the south began the wars that would lead to an Italy united in 1870. Mazzini realized his dream, though others played parts that an exile could not perform.

Mazzini wrote a great deal, but he was not a systematic thinker who would develop a fully coherent philosophy in the manner of many nineteenth century thinkers, including Mill, Marx, and Darwin in the England in which all four resided. Mazzini's writings hark back to the Romantic era and are in the tradition of the idealist position out of which early nationalist philosophy developed.

As well, his work sometimes looks forward. His appeal is often emotional, presaging the twentieth century when politics began to be seen as far different from a simple rational calculation of what is in one's best interest. The nationalism he espoused cut quickly to matters of identity, community, desire, and emotion.

Mazzini's most important work came to be *The Duties of Man*, published in early 1860, a summary of many of his ideas. He addressed the essay to the Italian working class, appealing, as he always did, to those left out of modernity, in the belief that nationalism would include a reconstruction of class and economic matters as it ended the power of the various *ancien régimes* still present on the European continent. In this matter, he and Turgenev are close, for both believed that serious reform and transformation were necessary in their countries.

In this work and others, Mazzini invokes God:

> The origin of your duties is in God. The definition of your duties is found in His law. The progressive discovery and the application of His law is the task of Humanity.
>
> God exists. I do not need nor do I wish to prove it to you; to try to do so would seem to me blasphemy, as to deny it would seem foolishness. God exists, because we exist. God lives in our conscience, in the conscience of Humanity, in the universe which surrounds us.

Mazzini's God is various, depending upon the context in which it is discussed, again showing that he does not worry about developing a self-referential philosophical system. Often God is discussed in a manner akin to the eighteenth century deists, sometimes even combined with the romantic concept of theism. God for him represents something beyond our own will which binds us to the universe and to fellow human beings.

On occasion, Mazzini will discuss God in the context of Christianity, for he and all other Italians at the time needed to deal with the power and presence of the papacy and the local churches. When he does so, he generally does not deal with Roman Catholic theology, but with the importance of following what he terms the will of God. It is not a Christ-centered discussion at all, though Jesus is on occasion invoked as a model.

For Mazzini, God is on the side of the people, not owned or necessarily represented by any institution or person. For him, the idea of the people is very close to the German concept of the *Volk*, invoked by such figures as Fichte and Hegel in the generation before Mazzini, as modern nationalism was first articulated:

> Preach in the name of God. The learned will smile; ask the learned what they have done for their country. The priests will excommunicate you; say to the priests that you know God better than all of them together do, and that between God and His law you have no need of any intermediary. The people will understand you, and repeat with you: *We believe in God the Father, who is Intelligence and Love, Creator and Teacher of Humanity.* And in this saying you and the people will conquer.

If nothing else, and important in the context of Italian thought at the time, Mazzini is ending the monopoly of the Roman Church on the concept of God and the practice of religion. In the end, the Church would be far weaker as a result of Mazzini's efforts to unite Italy and it will not recognize the legitimacy of the new Italian state until its pact with Mussolini in 1929. Hence, in practical terms, Mazzini's deference to the concept of God permitted Italian nationalists to follow their philosophy without feeling that they were on the side of atheism and immorality, no small accomplishment.

Another important concept repeatedly cited by Mazzini is that of Humanity. Like Hegel, Mazzini takes the long view. Individuals die, he notes, but Humanity (nearly always capitalized in the English translation) continues. Hence, we have a responsibility to the collective and to its development and progress:

> We improve with the improvement of the Humanity; nor without the improvement of the whole can you hope that your own moral and material conditions will improve. Generally

speaking, you cannot, even if you would, separate your life from that of Humanity; you live in it, by it, for it.

One's duty to the whole comes first, even before our responsibility to our family.

Here and elsewhere Mazzini asserts his belief in that nineteenth century concept inherited from the Enlightenment, the idea of progress. He uses it to abandon the notion that the purpose of life is to achieve salvation in the next world, to challenge, especially in the context of Italy, passivity and suffering. "The earth is not a place of *expiation*," he claims, "it is the place for working the *ideal* of truth and justice each of us bears implanted in his soul."

Mazzini is thus the heir of what Carl Becker called the heavenly city of the eighteenth-century philosophers, the world-view of Condorcet who predicted ever-growing progress. Indeed, he is Platonic in his belief, calling for a movement on a "ladder of perfection" whereby, by action, we will benefit humanity.

He, too, along with Marx, Turgenev's Bazarov, Mill in his defence of liberty, and others, is a philosopher who wants to change the world. And after his discussion of God and Humanity, Mazzini moves to show how he believes people must act in 1859.

Your nation is the field of action. For Mazzini, the individual is too small and lacks an identity without belonging to a community. Humanity is too large to try to organize without an intermediary body. It is what he calls the country that is the new important entity. He calls on God to justify his ontology—it is God who divided the world into different groups and thus "planted the seeds of nations." However, in Europe, which is what interests him—he never discusses the rest of the globe—the sovereignties are bad ones, both in terms of social morality and in the artificial divisions into dynasties and families in power. Simply, he predicts, "the map of Europe will be remade."

Hence, he too asserts the reality of the nation. Without a country, he argues, you have no identity, no agency, no real being, no rights. Your country is your home, you labor in it for your benefit and for the benefit of all. We Italians must exist as an independent country if we are to realize our potential as individuals and as a nation.

Mazzini's appeal goes beyond rationality. It is idealistic and sentimental. It reifies the idea that our identity is ethnic. Who are you?, he asks in his appeal to the people of Italy. Your answer: "I am an Italian."

Mazzini insists that a country must be unified as one. Hence, his Italy extends geographically from the tip of Sicily to the northern areas of Piedmont, Lombardy, and Venetia. He opposes any serious federalism, insisting that Italy must be single. There is, he claims, something of a "national tradition" to which all must relate.

Critics of Mazzini, including those who sympathize with his work and ideas, often note, quite rightly, that his Italy was ideational rather than actual. What he called Italy was a group of local entities, each with its own dialect and history. At the time he is writing there is no such thing as a national Italian language, and what he called Italy had not been united since the fifth century. But those in the 1860s who insisted that the nation was a reality—from the Germanies to Italy to the United States—ignored reality in favor of an idea and a teleology.

Mazzini had no hesitation in acknowledging the feelings of being part of a nation: "A country is not a mere territory. The country is the idea…; it is the sentiment of love, the sense of fellowship which binds together all… of that territory." A country has a mission, one more powerful than that of the Roman Empire or Roman Church. Using a literary and artistic term, he calls the idea and goal "sublime."

In the context of 1859, Mazzini's ideas are radical not only in his seeing a need to transform the map of the continent. There is, as well, a political and social goal attached to the change, one left over from the revolutions of 1848. He takes the idea of a country further. It is "a fellowship of free and equal men bound together in a brotherly concord of labor towards a single end."

Hence, he grafts liberal ends onto the national purpose. Individuals in a country must have rights; privilege based upon birth and tradition must end. Law must be made in concert, and no small group of citizens should control politics. Class must be set aside in favor of association.

As nationalism would evolve, especially ethnic nationalism of the variety preached by Mazzini and others, the doctrine would align itself with force, authority, and inequality. The imperatives of the nation would be placed before the rights of human beings. The crimes committed in the name of national development would make the twentieth century one which opened with wars of national interest, witnessed the Bloodlands of eastern Europe from the 1930s to the end of World War II, and saw the twentieth century in Europe end with the ethnic cleansing crimes of the Balkans.

However, in the mid-nineteenth century nationalism could be found on the liberal side of the political spectrum, on the side of change and rights in the face of a multitude of *ancien régimes*. Logically, liberalism, cosmopolitan and emphasizing the free individual, and nationalism, communal and emphasizing identity as ethnic, could never fully unite. Where nationalism is liberal today, it is in the form of civic nationalism, not the ethnic variety. But in 1859, the two did have a common enemy—the reactionary regimes in much of middle and eastern Europe —and both were important agents of change. Later, they would part company.

Hence, Mazzini believed that a republic which was democratic was the only proper form of government. All citizens must be equal and all must share in making laws. Association means that the amorphous entity known as the people is sovereign. Here, Mazzini is again challenging the tradition of government by a class or a group of oligarchs. For him, as for the revolutionaries of 1848, states and governments must be re-constituted to reflect the moral principle that citizenship creates a society of equals.

Attached to democracy and a republic is his concept of liberty:

> Personal liberty; liberty of locomotion; liberty of religious belief; liberty of opinion on all subjects; liberty of expressing opinion through the press or by any peaceful method; liberty of association...; liberty of trade: these are all things which no one may take away from—except on certain rare occasions which it is not necessary to mention now—without grave injustice, without arousing in you the duty to protest.

Mazzini adopts the idea of natural rights as the basis of a moral civic society. Not even the People (his capital) may take away your personal liberty. It is here that Mazzini fits into the kind of thinking supported by his admirer John Stuart Mill.

As well, he follows Rousseau in arguing that individuals—or even a majority—cannot alienate anyone's liberty. Liberty is part of being human and being able to make choices about one's life.

Liberty, for Mazzini, is something that he attaches to the sacred, to being "under the ruling influence of the Idea of Duty and of Faith in the common perfectibility." The appeal is coherent and rhetorically persuasive. Yet it finesses issues where liberty and the notion that one's personality is part of a group, a nation, might conflict. Rousseau tied liberty to his concept of the General Will. Mazzini is not clear about this, but his emphasis on the nation and association seems to do so as well.

Mazzini recognized that his philosophy required nations to be involved in the education of their citizens. He first made a distinction between instruction and education. Instruction is about learning those things one needs to know to live reasonably—reading, writing, numeracy, etc. It appeals, says Mazzini, to the intellectual side of our nature.

What he calls education is that which appeals to our "moral faculties." Hence, it is education that teaches about association, the well-being of the individual in society. It turns instruction from something that is personal and that can be used for tyranny and authority into a means of securing progress in the advance of civilization and what Mazzini means by liberty.

He claims that the moral teaching of his time is a kind of anarchy. Left to the home it simply does not exist in areas of poverty where parents work so hard that they have little time or strength to teach their children. The other classes often imbue a materialism and/or a superstition that does not forward liberty.

The answer: "...National Education, from which alone a national conscience can issue." It is a means of giving the nation a moral foundation.

The nation, claims Mazzini, must establish a program of free compulsory national education. This will include a study of history, most especially national history, the moral principles which guide the country, and instruction in literacy and other matters. This instruction is not neutral: "every citizen ought to learn in these schools equality and love."

Mazzini saw national education as promoting liberty and rights. It will, he expected, form the basis for social progress in what he termed "the direction of the Good." Provide education and people will no longer be "slaves," those who are subservient to others. It is a formidable force against any tyranny.

The issue of national education was an important one in 1859. Indeed, it remains important today. Mazzini does not discuss the Church in the context of education, but in Italy and elsewhere in Europe education became a battleground between churches and governments, a source of great conflict in the social and political life of everyone.

Those who favored national education saw the new modern state as the source of identity and morality. It is national education that gave impetus to the establishment of national languages in the modern period and to the celebration of the modern state as prime in issues of identity and loyalty. As Eric Hobsbawm and others have pointed out, nationalism brought with it the "invention of tradition"—the celebration of new holidays based on the founding of the modern nation, monuments to heroes of the nation, and an educational system which socialized children into loyalty to the state.

Soon, modern states would have national days as legal holidays. They would battle religious leaders to gain full control of the education system, as France did in 1905. Their capitals would, like Washington or London, have new holy places, shrines to important national figures, statues to politicians and war heroes. Modern shrines are different from those of Europe's past. Now they are museums, filled with art that crowds come to in order to view their national and military history. The new important tomb is that of the Unknown Soldier, not a religious saint. Statues of dictators, presidents, and prime ministers have replaced those of saints. National anthems begin school days for children, not prayer.

There were major figures in 1859, sympathetic to what Mazzini was doing, who worried about national education. The most notable was John Stuart Mill. He recognized the need for universal education, but was puzzled about how to do it. National education, he thought, might well become socialization and conformity, a version of the tyranny of the state over individual development, a means of enforcing the will of the state rather than helping individuals to choose how to develop their best selves in line with their tastes and abilities.

Mill was correct in his concern, though he lost in the development of the modern state. National education, whether in democratic France or authoritarian Russia, remains the major way states inculcate loyalty and identity. It was a famous Jesuit axiom which said: "Give me a child until he is seven, and I will give you the man." Now the state has taken over from the Jesuits and other religious denominations. Here Mazzini may have been right or wrong, but he certainly was prophetic.

Mazzini did not ignore another main issue of the day: the economy and how it is to be understood and organized. He first focused on the reality of the times and the poverty of ordinary people. In addressing Italians he noted that three-quarters of males in the working class find that life is difficult, a struggle to find the means of existence.

Like the socialists and some liberals of the day, Mazzini was outraged at the distribution of wealth and found his current reality immoral. He defined his ideal:

> *Whoever is willing to give for the good of all that much of work of which he is capable ought to obtain enough recompense to enable him to develop his own special life more or less in all the aspects which define it as human.* (his italics)

The dilemma is that poverty was systemic in Italy and the Europe of his day. The result, an analysis that could have come from Marx, Bakunin, Proudhon, and others of the time, is that most people must spend all their efforts simply meeting basic needs. They cannot progress in such an environment.

It is not production which is the problem, it is distribution. Free competition leads to a kind of tyranny on the part of those who have capital. Indeed, capital in his time, states Mazzini, is "the curse of... economic society [and] the despot of labor."

At this moment Mazzini seems to be moving to a socialist position in his analysis. However, he quickly veers away from the conclusion of most socialists. Property cannot and should not be abolished, he claims. Rather, it needs to be regulated and redistributed so that many can acquire it.

Mazzini clearly distinguishes his ideas from those of communism and anarchism. Anarchy is in error in trying to abolish society itself, in simply leaving individual liberty as the goal, he states. Actually, some anarchists at the time were rebelling against the vastness and impersonality of large government, preferring cooperative living on a small scale as a way people could have control over their own lives.

Communism, the abolition of private property, he states, will never reach its ends using those means. Here Mazzini is correct in arguing that the communism of the day will not produce equality of labor, nor will it result in a fair distribution of goods. The state will not wither away—it will make decisions about the lives of people in a manner that will usurp the very freedom it claims to be offering.

The proper goal is to put capital and labor in "the same hands." Everyone needs to be both a producer and a consumer and there will then be a fair society with an end to systemic poverty.

Mazzini calls his idea the "association of labor." The appeal is as much emotional as it is rational: "You were *slaves* once; then *serfs*; then *wage-earners*; before long you shall be, if you will it, free producers and brothers in *Association*." Through Association people are to be equal, to receive fair compensation for their labor, to be part of a brotherhood, to own capital collectively in addition to being a laborer. Mazzini's Association inside a national state looks very much like the socialism he claims to be rejecting.

The state has a role. Mazzini suggests that an actual national government, a republic of free people inside an ethical nation-state, will be able offer the assistance needed to reach the goal of Association. Assistance includes supporting associations, moral support via the legislature and the educational system, helping to build a substructure for communication and trade, encouraging public works to be done by associations, a simplification of unnecessarily cumbersome legal systems, and a new coherent system of taxation based on income tax.

Mazzini envisages a time when a new national state will appropriate the lands and possessions of the Church, as happened in France during its revolution. The state should reclaim land not used for cultivation. It will centralize and administer the railway system and other large enterprises. Wealth gained by the state will be put into a fund in order to provide cheap credit for development. A system of local government banks will be developed under the supervision of the central authority.

In summing up Mazzini again invokes his concept of God and uses language which is spiritual: "Three things are sacred: Tradition, Progress, Association." No individual can flourish or live a moral life alone. He implores the workers of Italy to contribute to the freedom and unification of their country as the only way to relieve their difficult conditions and end what he regards as their present social servitude: "...[T]he chief and most essential duty of all is to your Country."

For Mazzini, nationalism is the modern ideology which replaces that of traditional Christianity. In some ways he was enormously insightful. His nationalism is an appeal to identity, to the end of alienation, and to integrating the lower classes into the political framework.

He certainly was correct in believing that nationalism was the political stance which would dominate modern European history. States of various political ideologies all adopted a nationalist posture. Democracies like France and Britain; fascist states such as Germany, Italy, Spain, Hungary, and Portugal; so-called communist entities like the Soviet Union and its satellites in eastern Europe after World War II; breakaway entities such as Estonia, Latvia, Lithuania, and the Ukraine; reclaimed states such as Poland; and Balkan states such as Serbia, Bosnia, and Montenegro, all appealed to national identity as the basis for their legitimacy. For good or ill, they used national pride to develop diplomatic goals, from the *revanche* of France after the Franco-Prussian War, to the *lebensraum* and *Judenrein* of Nazi Germany, to the pitifully foolish imperial moves in Africa by Mussolini's Italy, to the new Pan-Slavism of Putin's Russia.

Hence, Mazzini's focus on nationalism as central to modernity meant that he was a better predictor of the future than Marx and most socialists, or conservatives who championed *ancien régimes*.

Mazzini's appeal to emotions and feelings, to the desire for belonging and even to the unconscious, was also very modern. He is probably the first major political thinker in modernity who did not attempt to build a coherent rational system. He certainly made little appeal to empirical data; he cared little for logical consistency.

Nationalism, in Mazzini's hands, became part of the quest for identity and its appeal, at bottom, was to sentiment and blood rather than to rights, utility, or an empirical understanding of historical development. Of course, in the twentieth century such would be the case in most national movements. The politics of irrationality and belonging often replaced those of natural rights and tolerance.

Mazzini's work is important in the development in modernity of both movements of national liberation and the growth of fascism. Movements of national liberation came from both the left and the right, as nationalism came to be neutral in its political orientation, something that all could incorporate into their ideologies and goals. What now became the best defence of the legitimacy of a movement or a state? Not dynastic control, not custom, not traditional geographic boundaries. Now it was the incorporation of the nation into a political entity which gave credence to borders and constitutions.

Mazzini's ideas of association and duty helped to make his work something of a precursor to the fascism of the twentieth and twenty-first centuries. It is unfair and intellectually irresponsible to blame Mazzini for the rise of Mussolini, just as it is unfair to accuse Hegel of being responsible for Hitler or Marx for Lenin or St. Thomas for Pius IX, the pope who would in his *Syllabus of Modern Errors* (1864) list, among others, rationalism, nationalism, liberalism, and socialism. But the importance of the state, of being part of the state by means of labor and association, is something that would be used by authoritarian regimes.

Mazzini's nationalism was, in the end, ethnic. While he insisted on the state as one which would provide rights and be a republic, the criteria for belonging was one's language and ancestry. He expected, as did others in the more innocent time before the Great War, that a Europe organized along national lines, a re-made map, would evolve into a peaceful continent. He even on occasion expressed the hope for a United States of Europe as the goal.

Ethnic nationalism came to represent a nasty turn in the political development of Europe. Most states on the continent have still not sorted out how to deal with the Other in their midst, especially the Other who is a citizen. Those who try to deal with this issue attempt to turn what was ethnic nationalism into civic nationalism, an insistence that all agree on a civic code separate from both religious belief and ethnic origin.

Still, in France, Germany, Sweden, and other Scandinavian states, and the Netherlands, the practice of civic nationalism in theory has not always been as civil as most would like it to be. It certainly has not erased strong appeals to ethnicity, so strong that political parties have grown on the basis of opposition to immigration and a veiled hatred of the Other.

Like many prophets, including his contemporary and fellow exile Marx, Mazzini erred in his vision of the future. He was correct in seeing nationalism, a national appeal based on emotion and identity, as the new way most Europeans would think of themselves. After all, even in Turgenev's two 1859 novels, nationalism plays a strong role in understanding desire, motivation, and identity. Sadly, he erred in thinking that the new map of Europe would result in contentment and peace. We know that vicious crimes, even new inquisitions, have been carried out in the name of national goals.

Interestingly, there came to be a kind of political battle, one still being played out in Europe, out of the reflections of three of the major 1859 figures, all of whom lived in or near London: Mazzini, Marx, and Mill.

All three resonate today in virtually every discussion about where Europe is heading and in the politics of its states. Nationalism remains a dominant idea, perhaps the dominant idea in relationship to identity, culture, and the relationship to a larger community. It is no exaggeration to consider nationalism as having replaced traditional religion in its role in society, education, devotion, and the understanding of the self.

However, Europe and the West—the world, really—has been transformed by socialist ideas, those central to Marx. Western, northern, and much of southern European entities are socialist, be they democratic socialists or social democrats. All accept, however imperfectly, a universal system of education, a health infrastructure open to all, the rights of labor, and various other goals of socialist thought. In the east, most notably in Russia where the Leninist-Stalinist version of the Marxist experiment was attempted, those goals are articulated, though in a system which more nearly resembles klepocratic fascism rather than anything resembling socialism.

Mill's devotion to liberty and to the importance of individuals being able to choose their modes of existence never goes away. Moreover, though he used both utilitarian thought and natural rights as a base, the natural rights tradition begun in France in 1789 has spread to most of Europe. Mill was a democrat, but one who feared some of its possibilities as he was writing at the time. He worried that democracy and universal education organized by the state might result in conformity and mediocrity. As he aged, he abandoned any interest in laissez-faire and moved to a social democratic position. He was a loyal Englishman and after 1859 sat in parliament for a time. However, though he liked and admired Mazzini, Mill never moved to nationalism as the answer to the woes of the time.

And we must not forget Turgenev's consideration of Russian nationalism and its relevance to Russia today. One wonders whether a Bazarov today might also state to the masters of Russia that much needs to be destroyed in a clearing of the air to move to an honest, open, and meaningful future.

Mazzini has on occasion been styled a "man of words" by commentators, some trying to put limits on his influence. They are wrong. For him words were acts, and words were what he could contribute to the making of his beloved unified Italy. In his fine study of Mazzini's ideas, Salvemini rightly claims that it was Mazzini who kept the idea alive, who was single-minded in supporting his goal, and who "was responsible for the psychological preparation" which brought about the kind of Italy that came into being from 1859 to 1870.[32]

More than that, he was the prophet of modern nationalism in its practice in Europe as a whole. His words became a new frame of reference for transforming the continent and elsewhere, no small accomplishment. That his legacy is both liberating and confining is to be laid at the feet of his successors.

---

[32] G. Salvemini, *Mazzini* (Stanford University Press, 1957), 158.

Chapter III
Ivan Turgenev: Russia on the Eve

Of all the great Russian authors of the nineteenth century—and there are many, including Gogol, Pushkin, Dostoyevsky, Tolstoy, and Chekov—Ivan Turgenev has come to be seen as the most Western. He lived for a time in Germany and in France, became friends with such contemporaries as Flaubert and Hugo, and wrote in a style that owes as much to the traditional Western novel as it does to those masters in Russia who preceded him.

Still, Turgenev was very much a "Russian" writer. His main works deal with the social and political conditions of Russia in his time and had much support and criticism at home. Like many Russian intellectuals—Westerner and Slavophile alike—Turgenev viewed Russia as part of Europe, and recognized that Russia was an underdeveloped country in both economic and social matters.

Moreover, Turgenev's main works, including his two novels of 1859—*On the Eve*, written in 1859, and *Fathers and Sons*, taking place in 1859 though written in 1860 and 1861—were very much about human relations and the deepest of human emotions. *Fathers and Sons*, among the most controversial works in all of Russian literature, has in its plot the stories of five love relations. Even so, in this short work, about 220 pages, Turgenev manages to say much about what has come to be known as the condition of Russia. He does so at a time when the whole of the ruling and governing classes of Russia, as well as the intelligentsia, were debating the reform of government and the end of serfdom.

The "condition of Russia" in relationship to itself and the West was part of Russian consciousness since the time of Peter the Great (ruled 1682–1725). It was Peter who believed that Russia, to preserve its autonomy and in order to develop, must begin to emulate the West. He imported scientists and engineers; he went so far as to put a tax on beards; and he founded that great artificial Euclidean new capital, St. Petersburg, as his "window on the West," facing Europe.

From that moment on, a main issue in Russia was its relations with others and its status as a "great power" at a time when a piece of the small European continent came to rule much of the world.

Russians were often ambivalent—they wanted the science and development of the West and at the same time declared their uniqueness as a people and culture. When they did supposedly emulate the West—for example in the reign of Catherine the Great (1762–1796), who liked to think she belonged to the group of enlightened despots of her time—the enlightenment really stopped at the border. Little was done internally to resemble Western society.

The many contradictions and paradoxes in Russian history from the time of Peter to 1859 were summed up in what was called the Westerner-Slavophile controversy. Westerners wanted Russia to become like the Germanies or France or Britain. Russian aristocrats often sent their children to schools in these countries, imported Western scientists and philosophers, and among themselves would speak French and/or regularly use English, French, and German phrases in their discourse to show their sophistication. Russia was backward, they claimed, and it needed to develop in the Western mode both to be able to compete internationally and to better the conditions of a serf society.

Slavophiles stressed the differences between Russia and the West and valued Russia's traditions and unique culture. They talked about the mystique of the land, the emphasis on communal life rather than individualism, the understanding of human nature as deeper and more spiritual than that of the rational man of the Western Enlightenment, the significance of the Russian Church and its highly emotional religiosity as opposed to the highly rational Jesuitical formulas of Roman Catholicism.

And, indeed, all acknowledged that Russia's historical experience differed significantly from that of the West. Russia did not have a Renaissance or Reformation; it did not develop a mercantile and capitalist society; it never valued the individualism so cherished by such figures as John Stuart Mill; and it did not import the ideas of the French Revolution. In fact, Russia's revolutionary tradition was nearly none—it looked to a small uprising of people in St. Petersburg in 1825, sometimes grandly called the Decembrist Revolt.

There was also the fact that, unlike much of the West, the old regime of Russia was still in place in 1859. The Tsar was both a political authority and a ruler who had a mystical presence. The Russian Orthodox Church was closely tied to the state and was one of its institutional and emotional supporters. A large number of Russians were serfs, tied to the land, governed by a gentry which was hereditary. Two main cities existed, Moscow and St. Petersburg, but at that time Russia was still untouched by the industrial revolution with the exception of some developments in the military. No bourgeoisie existed to act as the agents of change.

Many Russian intellectuals, Turgenev among them, adopted parts of both the Westerner and Slavophile postures in their perception of what Russia needed to do. They wanted development, they looked to the amelioration of the lot of the serfs, even some form of emancipation, but they also wanted Russia to be Russia, not to lose its unique sense of the land and its emphasis on communal life. They wanted reform, but did not want to become a larger version of France or Britain.

When reform was mentioned in the first half of the nineteenth century, the Tsars and their many powerful supporters resisted. They often pointed out that Russia, in its vastness and in its straddling both Europe and Asia, was unique. Moreover, they suggested, Russia required a different form of political and social structure than that of the West.

When challenged, the defenders of the existing Tsarist structure pointed to what they called the Patriotic War of 1812 as the moment of testing and as proof that Russia could compete and defend itself. Napoleon's invasion with roughly 600,000 troops was and is still viewed as an important moment in Russian history, a watershed that proved Russia was doing things correctly by its own standards.

As we know, Napoleon's defeat in 1812 marked the beginning of his downfall. But for Russians it meant that they were now a world power, taking an important place at the Congress of Vienna of 1814–15. As well, as has often been pointed out, it was not Russians who defeated the imperial Napoleon, it was Russia. The strategy of retreat, of drawing the French into the heart of Russia and into the Russian winter, succeeded. The myth of mother Russia and the land gained an even greater hold on its people, later celebrated in the great Tolstoy novel *War and Peace* (1869) and in Tchaikovsky's *1812 Overture* (1880), among other works of literature and music.

However, events conspired to make the years 1859–1861 very important for Russia and its rulers.

There had been no conflicts between major European powers since 1815, but this relative calm was broken in late 1853 by a war between Russia and Turkey which drew in other states. It began as a dispute over the area now known as Romania, and extended to the Balkans, then under the control of the Ottoman Empire. Russia had been pursuing an expansionist policy since the seventeenth century, absorbing Slavic areas on its border. In the eighteenth century the policy had been generally successful in the west and north, but not in the south, where Russia was eager to secure an outlet to the Mediterranean Sea.

Russian ambitions in the south had been checked by Britain and France, neither of whom wanted naval competition in the Mediterranean. Hence, when war broke out, Britain and France joined Turkey in 1854 in order to halt Russia. The war did not last very long and was not very bloody. Russia mobilized badly and fought weakly and it was decisively defeated. In the peace treaty, Russia was prevented from sending warships through the Black Sea into the Mediterranean and agreed "to respect the independence and territorial integrity of the Ottoman Empire." For the first time in centuries, Russia lost a European war.

The defeat—combined with the death of the very conservative Czar Nicholas II in 1855—forced the Russian imperial court and the gentry to consider implementing reform. In effect, domestic change was sped up by the need to make certain the state could compete internationally.

Reform was discussed at the court of Alexander II (1855–1881) and among the landed classes. The main matter was the continuity of serfdom in Russia and the need to transform the many serfs into a peasant class as a way of contributing to modernization. The issue was not new. Many Russian landowners and aristocrats who absorbed Western ideas, including Slavophiles, had realized that something needed to be done about Russia remaining a feudal state into the middle of the nineteenth century.

The sorry performance of the Russian military in the Crimean War became the catalyst for some change. Following the death of his mother, Turgenev himself had in 1850 inherited about 2000 serfs, peasants tied to his lands. He found himself in an ambivalent position, as did many enlightened landholders. With reform in the air, in 1858 he made a settlement with most of his serfs, putting them on a wage system, giving them as property the land which each had tilled. Turgenev lost roughly one-fourth of his income, but like one of his characters in the novel *Fathers and Sons*, Nikolai Kirsanov, who represents the best of the old gentry, he felt much better about the new arrangements.[33]

Alexander II was no liberal, but he did know that the time had arrived for reform. "It is better to abolish serfage from above," he said to his nobility in 1856, "than to await the time when it will begin to abolish itself from below." On February 19, 1861, in a *ukase*, or edict, he decreed the end of serfdom in Russia. The serfs became peasants, subject to the authority of the government rather than to their local masters. It was a complicated and halfway reform, for the former serfs were left with debts and many burdens, and the aristocracy remained very strong and in possession of roughly half the land.

Turgenev titled his 1859 novel *On the Eve*, knowing along with the rest of the Russian intelligentsia that change was in the air and inevitable. 1861 was a watershed moment in Russian history and in the consciousness of those who would ask the eternal Russian question: "What is to be done?" By 1864, the legal system of Russia was being reformed along Western lines, and local organized government was being introduced.

Prior to 1859, Turgenev identified a character type which would resonate deeply among Russian thinkers and readers and become one of the major mythic figures in modern Russian, and soon in Western, culture. An early short story of 1850 was titled *The Diary of a Superfluous Man*.

---

33 Leonard Schapiro, *Turgenev: His Life and Times* (Oxford University Press, 1978), 139.

Chulkaturin's diary is written over a period of a dozen days, from the time the doctor tells him on March 20 he will shortly die to his death on April 1. He takes stock of his life and tells the story of his having been in love, hopelessly, sadly, and having been ignored in rejection.

But, he asks: "Is it really worth telling the story of my life? No, it's definitely not worth it." Why? Because as he looks back at his life he decides that he has been superfluous. "I've invented an excellent term... People are bad, good, clever, stupid, pleasant and unpleasant; but superfluous... no." He states that the world would be no different if he had not existed, that someone joked that he was a throwaway card. He could not find his place in life, possibly because he looked for it "in the wrong place." He was excessively analytical, frozen from action because of his own self-consciousness. He calls himself a fifth horse, "completely useless."

The Superfluous Man was recognized as a type produced by Russian society. A sensitive soul, overly conscious of all that is around him, with no ability to connect or act, feeling he belonged nowhere in a society where his consciousness had no fit. Rejected by society, he turns in upon himself, and sometimes wishes that he would be envied by those who do belong, while he retains some contempt for them at the same time.

Russia's superfluous men at that time were educated, but had nowhere to fit in the gentrified traditional society in which they lived. They wished to belong, but they were alien. They wanted to act, but had no place to put whatever energy they did have. They were born and educated as Russians, yet they were marginalized by others and by their own ineptitude. They had modernist temperaments in the midst of a feudal world.

The ultimate insult endured by Chulkaturin comes when he decides to act, to insult Prince N__ at a ball. The prince, in Chulkaturin's mind, is his rival in pursuit of the beautiful Liza, with whom he is in love, and who clearly both loves the prince and is faintly repelled by Chulkaturin. The prince challenges Chulkaturin to a duel when the latter refuses to withdraw his remarks. He does so almost casually, as if this situation is just ordinary.

The duel occurs the next day, with pistols. Chulkaturin fires first and wounds the prince in the head, the bullet passing through his cap. "It's nothing," said the prince. "If I've been struck in the head and not fallen down, it means it is just a scratch." Now it is the turn of the prince, who simply states, smiling, "it's over," and he fires in the air.

Chulkaturin feels morally destroyed, for now the prince would be seen as magnanimous and totally decent. Even more important, he realizes that he is nothing to the prince and that others who hear the story of the duel will think that he wasn't even worth being shot. "By his magnanimity," he said, "[the prince] had literally banged a coffin lid down on me."

Turgenev's invention of the superfluous man struck a chord in Russian literature. Critics looked back on Pushkin's *Eugene Onegin* (1825–1832) and Lermontov's Pechorin in *A Hero of Our Time* (1840) and saw them as belonging to this unhappy category whose ultimate representative was the title character in Goncharov's *Oblomov* (1859).

Turgenev's friend Dostoyevsky in his *Notes from Underground* (1864) takes the matter further. Underground Man is spiteful and angry; he too writes a diary; he has an incident at age 24 where he wants to get in a fight and possibly get thrown out of a window, but his antagonist simply lifts him aside as if he is nothing; he wants to challenge an army officer who thinks he owns the street and obsesses for years before taking the smallest of actions; and he too has a Liza, in his case a prostitute who both understands him and wants to comfort him, and he commits a very cruel act to destroy the possibility of any continuing relationship.

Superfluous Man had many successors in Russia and the West, including many of Russia's intelligentsia in the 1860s. They will be followed by Robert Musil's Man without Qualities, Beckett's Vladimir and Estregon eternally waiting for Godot, and Ralph Ellison's Invisible Man.

The title of *On the Eve*, Turgenev's novel written in 1859, was firmly political and social. It was clear to him by then that Russia was on the brink of major transformation, that it was entering the world of Becoming after having spent so many centuries in Being. The future was on the verge of being open.

The inspiration for the hero of the novel was given to him by a neighbor, Vassily Karatayev, who had written a manuscript which Turgenev regarded as unaccomplished but interesting. When he went off to fight in the Crimea, Karatayev gave the manuscript to Turgenev, hoping that the latter would make something of it. In 1880, Turgenev recalled about Karatayev's tale: "This love story was told with sincerity, though unskilfully." At the time he was writing other works and only thinking about *On the Eve:*

> The figure of the heroine Yelena, in those days still a new type in Russian life, was outlined quite clearly in my mind; but I lacked a hero, a person Yelena could give herself to in her still vague, though powerful, craving for freedom. After reading Karatayev's story, I exclaimed involuntarily: 'Here is the hero I was looking for!' There was not such a man to be found among Russians at that time.[34]

---

34 David Magarshack, *Turgenev, A Life* (Faber and Faber Ltd., 1954), 196–97.

The work is set in 1853 before and at the moment of the outbreak of the Crimean War, which Turgenev in 1859 recognized rightly as indeed "on the eve." It is one among many of Turgenev's works—the most notable of which would be his next novel *Fathers and Sons*. Set in 1859, *Fathers and Sons* combines domestic life, tales of longing and love, and matters political in a way no other author in Russia, and very few others in the West, have managed. Turgenev is certainly among the first writers to connect the personal and the political so clearly.

His heroine in *On the Eve*, Yelena Nikolayevna Stakhova, is among the strongest and most interesting female characters in the literature of the mid-nineteenth century. She is raised on an estate outside Moscow and when we meet her she is 20 years old. She is not conventionally beautiful, even ordinary in looks, but she has a passion and an energy that attract and interest others. She adored her parents as a young child, but realized as she grew up that her mother, a classic hypochondriac, needed more attention than did she, and her father, who had openly taken a mistress, had little to give her.

Yelena is determined in a way that most young women of that time would regard as inappropriate and excessive. She judges others with strong convictions: "weakness exasperated her, stupidity made her angry, falsehood she would not forgive… A person had only to lose her respect… and he ceased to exist for her." Life was not easy or simple.

Above all, Yelena, who read widely, is not satisfied with the world of books. She is a new kind of woman, one who—like many intellectuals in Europe who followed Marx's dictum that the purpose of philosophy was not merely to understand the world but to change it—wanted to act. She has care and empathy for those who suffer in her world, the poor, those ailing, those who go hungry. Her father, totally exasperated, begins to wonder about his daughter and even fears her temperament, "describing her as 'some kind of zealous female republican, God only knows in whose image!'"

By the time she is in her late teens, she is independent. She has no female friends and life is lonely, a kind of struggle, says Turgenev, using a simile many other writers later use to talk about interesting women encased in domesticity, longing to escape: "she struggled like a bird in a cage." She not only wants to act, she understands that life is not full without love. She knew not who might she love, for no one seems right. She sometimes becomes ill, and her inner being has a nameless longing which she cannot clearly identify. She is in the habit of spending time in the evening at the window of her room, looking out at the world, reflecting on the paradoxes of her own consciousness.

The two young men in her life and in the life of the family are Shubin, an artist, and Bersenev, a student philosopher. In a conversation with Bersenev about his time at Moscow University, Yelena asks if there were any special young people among his companions. No, replies Bersenov, "there was not a single remarkable fellow among us." He notes that Moscow University is now an ordinary place.

However, Bersenov qualifies his remarks. There was a "truly remarkable man." He is Dmitry Nikanorovich Insarov, a Bulgarian studying in Russia, whose purpose in life is the liberation of his nation. Here, among the nationalists in the Crimea, Turgenev found his hero in his neighbor's manuscript.

Insarov is described as intelligent, having lost his father and mother to the Turks, experiencing the injustice of the controlling authority. He is not a proud man, is not wealthy, but he is a person devoted to his cause. As it turns out, Insarov will be living for a time with Bersenov and he is introduced to Yelena.

The two begin to meet more frequently and find themselves attracted to one another. Yelena is impressed with Insarov's character and his devotion to his cause. At one point, Insarov tells her:

> ...you asked me if I loved my motherland. What else is there on the earth to love? Which alone is constant, above all doubts, impossible not to believe in after God? And when the motherland is in need of you... Take heed of this: the last muzhik, the last beggar in Bulgaria and I—we want the very same thing. We all of us have one single aim. Try to understand what certitude and strength that gives us!

The words could have come from Mazzini, so certain and determined. Yelena muses: "So you would not stay in Russia, not on any account?"

Yelena falls in love with Insarov, as she records sensitively and sweetly in her diary. And that love is reciprocated with commitment and emotion. Shubin understands that Yelena loves Insarov not only for himself but for his passionate determination in the cause of his people, something still missing from his own Russian companions. Yelena and Insarov secretly marry, violating all of the traditions surrounding courtship and family. She informs her parents, who are stunned. The two are soon to leave Russia to take up the Bulgarian nationalist cause.

The contrast made by Turgenev between the Insarovs of this world of 1853 and Russian society and political culture is summed up later in the novel in remarks made by Shubin about the intelligentsia in Russia and about Yelena's commitment to Insarov, a passage highly significant in the politics of the time, telling about the 1850s:

> We [Russians] have no one yet, there are no human beings, no matter where one looks, it's all small fry, rodents, little Hamlets, Samoyeds, or obscurity and the subterranean backwoods, or arrivistes, pourers of emptiness into vacuity, and sticks to beat drums with! Or else it's the other kind: studying themselves with shameful subtlety, constantly feeling the pulse of their every sensation and reporting to themselves: that, they say, is what I feel, that is what I think... No, if there were any worthwhile human beings among them,

this girl, this sensitive soul, would not be leaving us, would not be slipping away like a fish into water!... When will our time come? When will some human beings be born among us?

Shubin's companion prophetically remarks: wait for a time, they will come. Russia will indeed produce many people like Insarov in the next several decades.

However, at the time of their departure, in November 1853, Insarov is ill. Yelena knows that she is leaving Russia for a long while, possibly forever, as the couple set out on their personal and nationalist adventure. They first go to Vienna, where Insarov lies ill in bed, then to Venice in late March 1854 on their way to Bulgaria.

Their time in Venice is stunningly beautiful and achingly sad, like the city itself. Turgenev's description of the city will be succeeded by many others, including Thomas Mann and Daphne du Maurier, who will find Venice also both magical and tragic, as is the love story of Yelena and Insarov.

In Vienna they go to the opera to see a performance of Verdi's *La Traviata* (first performed in March 1853 at La Fenice in Venice). The death of the young heroine from consumption foreshadows Insarov's own youthful demise several days later from an aneurism and lung disease. He never reaches his beloved Bulgaria alive.

Yelena is now faced with her own future and makes a clear decision. She asks Rendich, the Dalmatian who was to take them by sea to Insarov's homeland, to carry her and the body across the Adriatic, to allow the burial to happen in Bulgaria. I can do that, with difficulty, says Rendich, but I don't know that I can bring you back. Yelena replies: there is no need. I do not intend to return. She will take up the cause of her husband.

Yelena wrote her mother a letter the day following the death of Insarov. It is a farewell: "You will not see me again." Insarov's land will be her new homeland, his cause her cause. She realizes she is entering an incoherent and chaotic world of rebellion and insurrection, but she is impelled to do some good in the world, even if it means courting death. She asks forgiveness from her mother: "It was not in my power to act otherwise. But return to Russia—why? What can one do in Russia?"

Turgenev adds a coda, like many other nineteenth century masters, in which he tells what happens to the characters after the end of the tale. However, he enters new ground with Yelena's story, for it is possibly the first of many open-ended female voyages of modern times. He notes that in the five years after the death of Insarov and Yelena's decision to join the Bulgarian partisans, there has been no news or trace of Yelena. Her father searched for her after the peace was made but learned nothing new. There were rumors of a woman and a coffin, but nothing of substance. Yelena may or may not still be alive; the reader has to fill in the aftertale.

The voyage of a woman, especially a Russian woman, had no precedent or form. Yelena is not like many of the sensitive rebels of her generation, not like Emma Bovary (1856) or Anna Karenina (1877) or Hedda Gabler (1891). If she dies, it is in the service of a cause and a love larger than herself.

In discussing Turgenev's most famous and influential novel, *Fathers and Sons*, which takes place in 1859, his excellent biographer David Magarshack remarked:

> [T]he novel occupies a unique place in literary history: never before or after has a work of fiction, which had nothing to do with politics, produced such an expression of political passions. It was, as Turgenev expressed it, like pouring oil on a flame.

Magarshack was correct about the explosion. The novel immediately became one of the most controversial works of Russian and Western literature, for it aroused strong feelings in all who read it, and still does. However, he was incorrect about the politics. The work, though ostensibly a set of love stories, both joyful and sad, was clearly about the Russia of Turgenev's day and its political and social reality. Here and in other works, including *On the Eve*, Turgenev is in the long tradition of Russian literature which deals with political matters, from Gogol to Chekov to Zamiatin to Pasternak and Solzhenitsyn. Indeed, only a page after making his remark that it had nothing to do with politics, Magarshack quotes Turgenev, who said: "My whole novel is directed against the nobility as the foremost class of Russian society." And, in his hero, Bazarov, of whom Turgenev remarked that he agreed with his views on everything except art, he recognized a new type that would resonate deeply from that time on in the history and culture of Russia.[35]

Fathers and Sons is even more contemporary than *On the Eve*. It opens in May 1859 at a coaching-inn, where Nikolai Petrovich Kirsanov, whose estate is 10 miles away, awaits the arrival of his son Arkady, who is returning from St. Petersburg, where he has just finished his university degree. Nikolai manages the estate on which also resides his brother Pavel, a former army officer who had a sad romantic episode and now is a melancholic man of leisure.

Pavel dresses immaculately in elegant English-style suits and perfumes his moustaches. He wears a fez, in the fashionable Turquoise style, a cravat, and a stiff collar. Like many of his generation his speech is filled with French and, on occasion German, phrases.

---

35 David Magarshack, *Turgenev, a Life* (London: Faber and Faber), 216–217.

Arkady has brought a guest, a fellow student, a bit older than himself, Yevgeny Vassilyich Bazarov, who Arkady clearly admires and follows. Bazarov is a young scientist who will be taking a medical degree. He is also described by Arkady to his uncle Pavel as a nihilist.

Turgenev did not introduce the term nihilism, but this novel made it part of the Russian and European discourse. What is a nihilist?, Pavel asks his nephew Arkady, noting that the term derives from the Latin *nihil*, meaning nothing. Does he believe in nothing?

Arkady tries to clarify. It describes, he says someone "who looks at everything critically." No principle is taken for granted, even if that belief is common and cherished. The uncle again asks: "[I]s that a good thing?" It can be, replies Arkady. But Pavel is contemptuous of this position. "We of the older generation think that without principles taken as you say on trust one cannot move an inch or draw a single breath." Indeed, he says, first there were Hegelians, now there are nihilists, as if it is just a fad that will pass. Now, he says, returning to his ordered life, it is time to drink my cocoa. Please page the servant and have it brought in.

Bazarov spends his days doing scientific experiments on the estate and bantering with the peasants and servants, who take to his ways, seeing that he treats everyone equally, even if he sometimes makes fun of them. They see him as one of their kind, not gentry at all. As well, he will on occasion help people on the estate medically.

Bazarov annoys Pavel by his very presence and manner. Indeed, when the four men are taking tea, Pavel asks Bazarov about his studies. Do you admire the "Teutons"? Yes, replies Bazarov, the Germans are in advance of we Russians in physics and science in general. Pavel becomes annoyed with Bazarov's manner. Bazarov is polite but not at all deferential to his elder who is a member of the gentry. The narrator remarks: "Bazarov's complete indifference exasperated [Pavel's] aristocratic nature." He was abrupt, almost contemptuous of Pavel in his tone. And, indeed, in Pavel's world all these nuances are very important, though they mean nothing to the younger Bazarov.

Pavel tries to annoy Bazarov but it doesn't work. I believe in things that are useful, the younger man explains, but he also cuts short what he calls Pavel's "cross-examination."

Your uncle is determined to be right, he says to Arkady, who tells Bazarov that he was curt and hurt Pavel's feelings. Bazarov replies: "Do you think I'm going to pander to these provincial aristocrats. Why, it's all personal vanity with them, the habit of being top dog and showing off." Besides, I've just found a rare specimen of a water-beetle. Let me show it to you. That is useful knowledge.

Pavel waits for a good moment to continue the discussion, to have a skirmish with the ill-mannered young doctor. He tells Nikolai that their generation should not give up easily.

The good moment occurs several days later at tea-time. Pavel cites the example of the English aristocracy. I am, he tells Bazarov, "a man of liberal views and devoted to progress." That is why I revere the aristocracy at which you sneer. Pavel uses the traditional argument of those in an old regime to defend his class and status. We provide standards, we have character, we have a devotion to duty. Moreover, he argues, you may ridicule my dress and my habits, but they reflect a sense of self-respect, important to a stable society.

Bazarov counters: "you say you respect yourself and you sit with your arms folded: what sort of benefit does that do the *bien public?* If you didn't respect yourself, you'd do just the same."

I am a man of principle, claims Pavel, I exist out of the history of Russia. What use are your logic and principles?, asks Bazarov. Abstractions don't help anyone in the Russia of today. Do your principles give people the bread they need when they are hungry?

How can you live without principles?, retorts Pavel. On what do you base your behavior? Bazarov gives his reply in the context of Russia in 1859, a position which will be part of Russian history going forward: "We base our conduct on what we recognize as useful. In these days the most useful thing we can do is to repudiate—and so we repudiate." There is nothing worth saving. This is not a time for construction: "the ground must be cleared first."

You recognize nothing, claims Pavel. You are not really Russian. An easy rejoinder for Bazarov: My grandfather, he tells Pavel, tilled the Russian soil, thus invoking the myth of the land in his support. Moreover, ask your peasants which of the two of us he would recognize as a fellow Russian. You will lose. You, in your aristocratic cocoon, don't even know how to converse with them.

The argument continues. Bazarov makes it clear that he believes that the Russian intelligentsia to the present has not accomplished anything. The coming emancipation of the serfs, in whatever form, will not solve the problems of the nation. He is no reformer or tinkerer. Russia needs to begin again.

Pavel loses his temper. What are you encouraging? You are like those we view as barbarians, the Kalmuks or the Mongolians. All good Russians stand for civilization, for our sacred beliefs and traditions. You will be crushed, you are only a few.

Perhaps we will be destroyed, says Bazarov. There are more of us than you think. You ask if we can take on the whole of Russia, but remember, citing history, a small candle managed to set the whole of Moscow on fire. I'll listen to you "when you can show me a single institution of contemporary life, private or public, which does not call for absolute and ruthless repudiation." Reflect on your classes and your society and let me know what should be revered. In the meanwhile, says Bazarov, ending the quarrel, I will do something useful and go and dissect frogs.

Turgenev, in Bazarov, has by 1861 found his Russian hero, the person he could not place early in the work about the previous decade. Bazarov is the new man, someone who will act, who has the qualities that Shubin in *On the Eve* said did not exist among the little men of 1853. He is no small rodent or an arriviste.

Bazarov has beliefs, wants a different Russia, and sees that its feudal structure needs to be destroyed and replaced eventually by something more decent and humane. His nihilism is very different from what nihilism has come to be in contemporary discourse. It is time to get rid of the old regime, even to simply destroy it and clear the ground. The reader might be reminded that one of the first acts of the National Assembly in France after the overthrow of the old regime was the law to end feudalism, even before passing the Declaration of the Rights of Man and the Citizen. Bazarov's nihilism is not at all Burkean, as Pavel would have his world be. It is a plea to rethink how his society should be reconstituted in light of humane considerations.

Bazarov has no Marxian illusions. He does not idealize the peasants or servants. He does know that the Pavels of his world are useless, that their continued existence simply means there will be no real change. He is dangerous because he cannot be co-opted by the existing elite. His intelligence and charisma are rightly seen as a challenge to traditional authority.

The dilemma Bazarov has in the year 1859 is that of the gap between his consciousness and the landscape of action. He is modern in his concerns and thoughts and wants change. However, the world in which he is living is still medieval. His consciousness is not in keeping with what is possible. He is ahead of his time.

In between writing *On the Eve* and *Fathers and Sons*, in January 1860, Turgenev gave a lecture titled *Hamlet and Don Quixote*, noting that the two great and influential works were written in the same year. He posited a dichotomy in talking about two kinds of human nature—Hamlet as self-possessed, reflective, and incapable of action; Don Quixote as an actor in the service of his ideals even in the face of imminent defeat. In Bazarov, Turgenev found neither the one nor the other. In many ways Bazarov is closer to a third modern myth, Goethe's Faust, than to the other two. He wants to act, using action as a form of consciousness and a demonstration of philosophy. He wants to help bring Russia into modernity. Unlike Faust, sadly, even tragically, Bazarov does not get that opportunity.

Critics have sometimes had difficulty with the complexity of Bazarov. He is not always consistent. A modern type, he is filled with paradox and contradiction. He is an empiricist who, surprising himself, falls in love. He seems rough, though he is very tender with children and the ill. He is passionate about his cause, but forgives those who do not join him. There is even a recklessness about some of his behavior, as if he enjoys courting danger unnecessarily.

But in Bazarov Turgenev understood something new was appearing in his beloved Russia. Bazarov is the precursor of the nihilists and *narodniks* of the 1860s and '70s in Russia, populists who wanted "to go the people" and both teach and foment dissent in the name of new ideals. The first organized Russian revolutionary party, *Narodnaya Volya*, The People's Will, secret and terrorist, was founded in 1879. And he is certainly the grandfather of the revolutionaries of the 1890s and the period from 1900–1917. Many of them took the position that Russia must indeed begin anew. Trotsky, Lenin, Bukharin, Radek, Zinoviev, and others owe their pedigree to the fictional nihilist of 1859. Bazarov is to Russia what Che Guevara—the myth after his death rather than the reality—has been to Latin America since the 1960s.

As strong and as dominant as is Bazarov, the novel is filled with other interesting and rounded characters. The main female is the "striking" (Bazarov's first reaction on seeing her) Anna Sergeyevna Odintsov. Anna is 29, a young widow from a marriage to a decent older gentleman. She inherited all of her husband's property and wealth, a circumstance which made her independent and in control of her own life. She is well read, somewhat distant from most people, determined and controlled.

Bazarov meets her at a ball—though Bazarov does not dance—and he is struck by her intelligence and manner. An empiricist, a man of facts, Bazarov surprises himself by falling in love after the two see one another several times at Anna's estate. The two undergo a transformation, as both, aloof, become more intimate with every encounter.

Finally, it is Bazarov who crosses the line into commitment. In an emotional encounter while visiting Anna at her estate, he confesses his love. She is deeply moved, but fearful. They embrace for a moment and then Anna moves to a corner of the room. "You have misunderstood me," she tells Bazarov, and he leaves the room. He sends a note saying he will be leaving.

Anna reflects: "I did not understand myself either." She blames herself for not foreseeing that the relationship had moved to love. She reflects on her confusion, for this is a key moment in her life, one which will determine which road she will travel. She decides she cannot make the leap—the leap that Turgenev's heroine Yelena made in *On the Eve*. No, she thinks, a quiet life is the one I desire.

She is very sad and, without cause, bursts into tears. The narrator tells us that she even has a sense of guilt. She has contradictory emotions—a sense that life could be exciting, a desire for something new and meaningful—but she cannot risk it. She cannot enter what the narrator tells us is a kind of "void... chaos without shape." For she realizes that Bazarov brings with him a life in which the future is open, and totally new, filled with danger.

Anna and Bazarov retain a relationship. When he is leaving she tells him: "We shall meet again, shall we not?" "As you command", he replies. "In that case, we shall." For each, the other is the most remarkable person they have known, the deepest emotional attachment they will have.

There are other interesting women. One is Fenichka, the young 23-year-old mistress of Nikolai, and the mother of their baby Mitka, half-brother to Arcady. Her mother was the housekeeper of the estate until her untimely death from cholera a few years earlier. Fenichka, able, wise in her own fashion, very kind and decent, modest, succeeds her mother. Nikolai's first wife—the love of his life—died suddenly and tragically in 1847, and her death leads him to the countryside and the management of his lands. His liaison with Fenichka is not at all casual, though it is unusual. They do not marry, for Nikolai worries about Pavel's reaction as well as that of his son Arcady.

On learning of his father's companion and that he has a half-brother, Arcady responds in the most generous and enlightened fashion. He wishes the couple well, he is delighted at his father's happiness, and he is excited about having a young sibling.

Of all the characters in her life, Fenichka is most at home with Bazarov. He becomes a doctor to her child, and they sometimes meet in the garden for a chat as the novel progresses. Her ease with him comes about because she senses that Bazarov has no intimidating aristocratic airs: "In her eyes he was both an excellent doctor and an ordinary man." And Bazarov finds he can easily relax in her presence.

Bazarov eventually goes to visit his parents, his father a retired army doctor, and his mother who has a small property in the countryside that provides the two of them with a modest and decent life. They are a Russian Baucis and Philemon—devoted to one another and their son, decent, kind, generous, living a quiet domestic life.

Turgenev is especially fond of the mother, Arina Vlassyevna, "a true Russian gentlewoman of the old school" who probably should have lived a few centuries earlier. It is a folk Russia that Turgenev finds authentic and meaningful. His discussion of Arina Vlassyevna should silence any critic who claims Turgenev was a simple Westerner.

Bazarov's mother is devout, superstitious, a believer in folk-tales and folk remedies, full of rituals to ward off bad spirits and illness. She is extremely kind, warm, loving, and loyal. She loves her son more than anything and hovers around him too much, worried that something evil might strike. She accepts without reflection the division of society into gentry and common folk, but is uncommonly gentle with those who are her servants. She is charitable to a fault and though she loves food, she always fulfills her religious duty of fasting. She rarely criticizes anyone, and has never embarrassed a soul. She defers to her husband in the running of their small estate, and worries when hearing the discussions of reforms. Though living a good life, she fears impending disaster every day. She both laughs and cries regularly. The narrator of the novel remarks: "Nowadays such women as she have ceased to exist. Heaven only knows whether this should be a matter for rejoicing!"

It is an incident around Fenichka that provides the occasion for Pavel and Bazarov to again clash. Bazarov is seen by Pavel to be too close to Fenichka. Indeed, it turns out that Pavel nurtures his own secret affection for his brother's companion. He insults Bazarov according to the aristocratic code—"I cannot endure you; I despise you"—and challenges him to a duel. Bazarov accepts, letting Pavel know he, too, can behave like a gentleman.

The encounter is both farce and possible tragedy at the same moment. After all, one of them can be seriously wounded or killed in this courtly ritual.

They decide on pistols. Pavel shoots first and a bullet is heard by Bazarov but not felt. Bazarov then shoots without taking aim. His bullet hits Pavel in the thigh. Pavel claims that each has one more shot, but Bazarov wisely ends the encounter by becoming a doctor and helping Pavel who faints. The wound is bloody but not at all serious.

Pavel is magnanimous and silently weary at the same time. Fenichka tells him of her love for Nikolai and Pavel finally realizes that his life has been wasted. Bazarov was correct in his judgment that Pavel harbored deep affection for Fenichka.

Pavel tells Nikolai to marry Fenichka and thereby ends the barrier for that to occur. He decides to go away, to live in exile in a place like Dresden or Florence, as did others of his sort. Lying on his pillow, recuperating, the narrator tells us that "his handsome emaciated head lay on the pillow like the head of a dead man." The narrator then remarks: "...indeed, to all intents and purposes, so he was."

Turgenev is correct, though in advance of the events. The leisured gentry class of Russia, like that of France and other old regimes, indeed are consigned to what Marx called "the dustbin of history" a few generations hence.

In his portrayal of the gentry, Turgenev took care to create characters who were good, who we might like and/or sympathize with as people, for he believed that this was a clearer way of proving that they needed to end, as what he called "a progressive class." Nikolai is a splendid and cultured person—he tries to manage his estate well, he looks after his peasants and servants. Like Turgenev, he frees his serfs before the ukase of 1861; he wants to marry Fenichka, thereby moving beyond the class barriers of the old Russia; he plays the role of local medic to the area; reads in science and literature; and even plays the cello with feeling.

In a letter of April 12, 1962 to K. K. Sluchevskii Turgenev noted:

> *My entire tale is directed against the gentry as a progressive class.* Take a look at the characters of Nikolai Petrovich, Pavel Petrovich and Arkady. Weakness and flabbiness or limitations. Aesthetic considerations made me take specifically *good* representatives of the gentry so as to prove my point all the more surely: if the cream is good, what does that imply about the milk?... [Nikolai, Pavel and Arkady] are the best of the gentry and that's precisely why I chose them, in order to demonstrate their bankruptcy.[36]

Turgenev—and many of his contemporaries who are what he called negators, including Belinsky (to whom the book was dedicated), Herzen, and Bakunin—are not reformers in the mode of the critics of the industrial revolution on the other end of the continent, in Britain. Benjamin Disraeli (*Sybil, or the Two Nations*, 1845), Elizabeth Gaskell (*Mary Barton*, 1848, *North and South*, 1854–55), and Charles Dickens (*Hard Times*, 1854) all critiqued British society and politics with regard to the conditions of labor and the poor in their time as terrible and immoral. But they argued for reform, not negation, not revolution. They hoped their novels would wake up the upper classes to their responsibilities. Not so in Russia. Turgenev here joins the plea for deep change. He is, as he noted often, on the side of Bazarov.

Arkady accompanies Bazarov in his visits to Anna Odintsov's estate—he is even for a time infatuated with Anna—and over time he becomes attracted to Anna's younger sister, Katerina, known as Katya, a charming, intelligent, well read, musically talented, perceptive young woman. She notes to Arkady that her sister was under the influence of Bazarov, "just as you were."

Arkady catches the use of the past tense and enquires about it. Do you see any signs, he asks, that his influence over me is waning? I do know that you never did feel sympathetic to him. Katya responds: "It's not that I don't like

---

[36] *The Essential Turgenev*, ed. Elizabeth Cheresh Allen (Northwestern University Press, 1994), 749–750.

him, but I feel I have no contact with him and I'm a stranger to him... just as you haven't anything in common with him either." Arkady seeks further clarification of this insight and Katya stumbles on a metaphorical explanation: "He's a wild beast, while you and I are domestic animals." Arkady is hurt by the remark at the time, though he will soon acknowledge that she is correct.

Arkady soon proposes marriage to Katya. Anna, Katya's older sister and surrogate mother, needs to give her assent, and she consults with Bazarov, who advises her that this will be a good marriage for all parties. Bazarov then informs Anna he will be leaving—he feels he has been at leisure for too long. They part, each knowing that a possibility was passed by.

Bazarov says good-bye to Arkady as well, congratulating him on his impending marriage, praising Katya's temperament and intelligence. But he tells Arkady that their friendship is also ending, as Turgenev uses the moment to distinguish between the decent gentry and the negators:

> And now, in parting, let me repeat... because there is no point in deceiving ourselves—we are parting for good, and you know that yourself... you have acted sensibly: you were not made for our bitter, harsh, lonely existence. There's no audacity in you, no venom: you've the fire and energy of youth, but that's not enough for our business. Your sort, the gentry, can never go farther than well-bred resignation or well-bred indignation, and that's futile. The likes of you, for instance won't stand up and fight—and yet you think yourselves fine fellows—but we insist on fighting. Yes, that's the trouble! Our dust would corrode your eyes, our mud would sully you, but in actual fact you aren't up to our level yet, you unconsciously admire yourself, you enjoy finding fault with yourself; but we've had enough of that—give us fresh victims! We must smash people! You are a nice lad; but you're too soft, a good little liberal gentleman...

Arkady is not put off by this speech, for he does understand his former mentor. The two men embrace and Bazarov, after entering the wagon that will take him away—who knows where in the end—calls out: "Farewell, brother!"

Bazarov returns to his parents' home, where he finds an occupation in helping his father, now so old that his hands shake, with his patients. And this turns out to be his undoing. While performing an autopsy on a typhus patient for the district doctor Bazarov cuts himself. The doctor had no silver nitrate to treat the wound and by the time Bazarov gets some from his father it is too late.

The illness comes and there is no remedy. Bazarov asks that Anna Odinstov be notified, and she comes immediately. "A royal gesture," Bazarov tells her. He handles his dying with great dignity, but with regret: there are matters to deal with in Russia, and now he will not be able to attend to them.

Later, in 1874, Turgenev called Bazarov "my favourite offspring."[37] To a young critic in 1862, he wrote: "My vision was of a great wild crepuscular figure, half emerging from the soil, powerful, angry, honourable, yet doomed to destruction because [he] stood, after all, only on the threshold of the future..."[38] He stated many times that if the reader does not admire Bazarov, then it is he, Turgenev, who is at fault.

No Russian author—perhaps no European author of his time—was able to combine the personal and the political as insightfully and as sympathetically as did Turgenev. *On the Eve* and *Fathers and Sons* can easily be read simply as tales about individuals dealing with their feelings and identities. Indeed a number of literary critics do so. However, the works are at the same time profound commentaries about what was happening in Russian society and culture, very important contributions to the debate on the condition and direction of Russia.

Russia modernized with difficulty and, in Tsarist times, reluctantly. In 1859, it was the Ottoman Empire which was styled "the sick man of Europe," though both the Russian empire and the Austrian empire could also easily qualify.

Bazarov was the new man of Russia, and he was rightly feared by the gentry and those who thought Russia could survive without modernizing to the point of creating a bourgeoisie. He became the model of the new intelligentsia of the underdeveloped world—educated, alienated, committed to the nation, determined to foment change in the name of a social ethic, even if it meant revolution, because reform was merely a way of propping up the old regime.

As we know, Tsardom fell in Russia from below and from within. The later Russian intelligentsia led the rebellious opposition, an opposition made up of a variety of ideologies and political alternatives. The Tsarist regime itself in the nineteenth and early twentieth centuries was hopelessly incompetent in times of crisis. It failed to do what it claimed was its divine mission and its justification—the defence of Russia and the protection and well-being of its peoples.

The new men of Russia in the first half of the twentieth century regularly claimed that Karl Marx was their spiritual father. However, in their behavior and their defence of the fatherland, it is Bazarov who was far more their guide.

Sadly, these days whatever Bazarovs do exist are either in jail or in exile. Russia has been highjacked by the secret service (the KGB) of the former Bolshevik regime and follows neither anything resembling what Marx envisioned nor what Turgenev and his Bazarov desired. There are no reformers or rebels in power, only thieves and thugs, nasty people who love power, ooze greed and,

---

37 Shapiro, op. cit., 185.
38 F. F. Seeley, *Turgenev: a reading of his fiction* (Cambridge University Press, 1991), 216.

in the name of nationalism, invade Ukraine. It is time for contemporary Russian literature to again rise in defence of the best of its land and culture. The eternal question remains: what is to be done?

Chapter IV
Jacob Burckhardt: Inventing the Renaissance

Jacob Burckhardt (1818–1889), a distinguished historian in and of Europe who had a professorship from 1858 at the University of Basel in Switzerland, his home town, said about his studies: "History is on every occasion the record of that which one age finds worthy of note in another."[39]

Burckhardt was unusual. At a time when the discipline of history was being formed and shaped, most practitioners of the craft were interested in politics and diplomacy. Indeed, Burckhardt left Basel in 1839 to study for four years at the new University of Berlin, founded in 1810, an institution which quickly became known for its innovative and creative professors and research.

Burckhardt studied with some of the most important historians of the time, most notably in the seminar of the one person regarded as the founder of history as a *wissenschaft*, a discipline, Leopold von Ranke (1795–1886). Ranke's most famous statement about the discipline was that he now wanted historians to tell the past *"wie es eigentlich gewesen,"* as it actually happened. No more romantic tales in the manner of Walter Scott, no more claiming to know about the past without looking at the documents. He emphasized scrupulous care in getting to sources and he and others developed the rules for the evaluation of the authenticity of sources and interpreting them which would influence all historians ever after.

The respect for political and diplomatic sources was in part a belief on Ranke's part and others that history was to deal with the nation and the development of the state, something akin to what Hegel in philosophy, Savigny in law, and Niebuhr in classical studies were also doing at Berlin. Hence, in its earliest phase, the discipline emphasized finding out the facts and recording national histories, sometimes political, sometimes military.

Burckhardt respected Ranke, and he did well as his student, but he didn't follow his lead. Rather, Burckhardt rejected the idea that power in the state reflected the vitality of the culture, as well as the emphasis on politics. Rather, he turned himself into a cultural historian with his first major book, a study of the world of fourth century Europe, *The Age of Constantine the Great* (1852).

There followed a trip to Italy which, like many Italian journeys of intellectuals and others in the nineteenth century, changed him and his perspective. Out of

---

39 Note: the edition of Burckhardt's *The Civilization of the Renaissance in Italy* used is: trans. S. G. C. Middlemore (New York: The Modern Library, Random House, 1954).

this came the *Cicerone, A Guide to the Works of Art of Italy* (1855), a work not only scholarly, but one used by travellers as well. He took notes on all his Italian journeys and thus produced a highly original work, both in drawing attention to neglected works as well as in interpretation. As he said: Italy provided him with "new standard for thousands of things." He saw the art as reflecting not only aesthetics, but the valuative side of life. As he said of Raphael: his highest "personal quality... was not aesthetic but moral."[40]

Burckhardt then spent time in Zurich researching his great master work, *The Civilization of the Renaissance in Italy*, published in 1860, a study which totally transformed how we think about Italy in the Renaissance and the meaning of the Renaissance itself.

As the discipline of history developed in the first half of the nineteenth century it became highly specialized and its narrative was almost always temporal, reciting one event after another—perhaps commenting on some matters—in chronological order. Indeed, in this development, historians specialized, as many commentators have remarked, as they began to be scholars who knew more and more about less and less. It reached the point where soon the boring, limited pedant/historian became a recognizable figure in literary works. One famous invention of this sort is the character of Edward Casaubon in George Eliot's *Middlemarch* (1871), a person of large ego and small mind. The classic character of the boring, banal, limited historian is found in Ibsen's *Hedda Gabler* (1891). George Tesman, Hedda's husband, spends his honeymoon in the archives, researching his contribution to scholarship, a study of the domestic industries in Brabant in the Middle Ages.

Not so for Burckhardt. His *Civilization* was not only a conceptual breakthrough of great importance in how we think about modernity. It was a cultural history which asked the reader to think holistically rather than consecutively. There are six parts, all looking at Italy in the Renaissance from a different perspective, all holding together like a work of art in the twentieth century. As well, it defined culture in the broadest possible manner. It was not only about art, literature, etc. It deals with politics in discussing the concept of the state; with human nature in defining a new personality type; with scholarship and history by relating how the past chosen as important changes the present; with discovery in the largest sense—that of discovering beauty and a new style of living in addition to journeys; with society and celebrations and gender relations; with ethics and the religious spirit in the life of peoples.

---

40 Introduction to *Force and Freedom*, ed. James Hastings Nichols (New York: Meridian Books, 1955), 10–11.

Ultimately it goes back to Burckhardt's statement about why we study certain times. They interest us because of who we are and how we understand ourselves. For Burckhardt, he had to re-evaluate what we call the Renaissance to understand how we moderns came to be who we are.

While the eighteenth century *philosophes* thought of themselves as enlightened and in a new historical period called the Enlightenment, the scholars and philosophers of the fifteenth and sixteenth century in much of Europe emphasized the revival of antiquity as the characteristic of their age. In the battle between the ancients and the moderns they deferred to their classical ancestors, unlike many in the Enlightenment and in the nineteenth century.

The founders of the discipline of history did accept that there was a periodization which recognized a distinction between the Middle Ages and something modern which could be recognized at about 1500. Otherwise, Ranke and his fellow scholars remained interested in politics and deep research into the archives of the state, acknowledging that something different was happening in the Renaissance in art and literature. For Hegel, the Renaissance was part of the dialectical process which led to the important German event of the Reformation. Its art was important in furthering the spirit of medieval religion, the revival of antiquity led to humanistic studies, and it led to an exploration of the world. All this contributed, thought Hegel, to an important development of the Spirit.

Jules Michelet, in his multi-volume *Histoire de France*, titled his seventh volume, on the sixteenth century, *La Renaissance* (1855). For Michelet, in the context of his view of French history, the Renaissance was not about Italy at all. It occurred in the sixteenth century, virtually spontaneously, and included the Reformation. He did use the term "the discovery of the world and the discovery of man." However, for him this was, as he put it, from Columbus to Copernicus to Galileo, and was understood in the context of French history. What he did initiate was the idea that there was something identifiable as the Renaissance and that it was different enough to merit being regarded as a distinct period of European history.[41]

It was Burckhardt who then defined the Renaissance for modernity and made it an important moment for those who wished to understand modernity's origins. And he moved it from France to Italy and to an earlier moment in history.

The Civilization of the Renaissance in Italy opens in an unexpected manner. Instead of beginning the story with the revival of the classics, Burckhardt chooses to talk about a new concept, "the state as a work of art."

---

[41] Wallace K. Ferguson, *The Renaissance in Historical Thought* (Cambridge, Mass.: The Riverside Press, 1948), 163, 173, 176–78.

There was the Holy Roman Empire and there was the Papacy and its lands. Between them, he notes, there were in Italy political entities which had left the feudal world, unlike what happened in France, and which also had no legal relationship to the empire. These are, in the fourteenth century, both republics and despotisms, formed recently. "In them," he claimed, "for the first time we detect the modern political spirit of Europe."

These sovereignties were governed in a new way, simply by the power of those who ruled to be able to maintain them. Burckhardt chose to focus on the despotisms. Their leaders had neither a birthright, nor an inheritance, nor the legitimacy found in simply being in power for a length of time. Rather, the State (capitalized in the translation and, naturally, in the German) is sometimes ruled by individuals who often are excessively egoistic or harsh. Still, in Italy in the fourteenth century and forward a new fact appears in history—the State as the outcome of reflection and calculation, the State as a work of art.

Burckhardt does not leave his Renaissance discussion to tell his readers why this is both new and important. Perhaps he assumed they would immediately understand the relationship of these created states to modernity. Once these Italian entities are ended, starting with the invasion of Italy by France in 1494, the notion of legitimacy in governance changes, ratified several times in diplomatic arrangements and sealed with the Treaty of Westphalia in 1648.

There came to be dynastic states all over Europe, different in their governance, all settled by tradition and custom. The English wound up with a dynastic state and an aristocratic parliament; the French had Louis XIV tell his people, "*L'état c'est moi*"; the Dutch had monarchs and guilds; the Russians had a Tsar. All had laws based on custom. Legitimacy was inherited.

However, modernity introduced something both new and, to Burckhardt, reflecting what had existed in Italy for two centuries. This is the idea motivating both the American and French Revolutions in the last part of the eighteenth century. Now, and it is the assumption behind every revolutionary movement and even some legal political ones, the state is not inherited, it is created. That is, the state from 1776/1789 can now be reconstituted, by persons or groups which take power and then rewrite the basis of law and governance. In modernity, Burckhardt, who himself was conservative (small "c"), is informing us that the state is not just something we inherit. It can be, and has been in his time, created. After all, his book was published as the new Italian Wars were being fought, leading in 1870 to the creation of the modern Italian state.

Burckhardt cites two individuals of the thirteenth century as exemplars of the new modernity. First, there is the Emperor Frederick II (1220–1250), who was the first to systematically attempt to destroy the feudal state, and who centralized political and juridical power in a way no one had yet attempted, for centralization is also part of modernity, as Tocqueville and others teach us.

Then there is Ezzelino da Romano, who married Frederick's 13-year-old daughter in 1236. Ezzelino wanted power and took it through war in Verona and other areas of northeast Italy. As Burckhardt remarks, Ezzelino stood for no special type of political organization or administration. Moreover, he was cruel beyond anything then known. But here, says Burckhardt, "for the first time the attempt was openly made to found a throne by wholesale murder and endless barbarities, in short of any means with a view to nothing but the end pursued." To seal the argument and to cover the whole of the period of the Renaissance, Burckhardt states that not even Cesare Borgia (1475–1507) committed such vile acts as did Ezzelino in the name of power.

These political entities in Italy, as they develop, are something new on the scene of Europe. They are what Germans call *Dinge an sich*, things-in-themselves. Burckhardt calls them the beginnings of "the purely modern fiction of the omnipotence of the State." It is an agreed upon "fiction" in the modern world, what epistemologists would call a convention, one very different from what had come before the Italian Renaissance and the development of the state. Sovereignty rests in states. They have the ability to make laws, organize systems of justice, print money, and make war and peace. He (it was always men in the Renaissance) who controls the state has a power and authority different from the individual citizen.

Hence, most of the despots behaved in an insecure manner, for they could be challenged by anyone, having no legitimacy beyond their own power and authority. How do you advance if that is the case? By "talent and calculation," for there is no objective legitimacy. Indeed, this new world was indifferent to the legitimacy of birth.

The sacred was a political and social matter in these dominions, not a religious one. Many had contempt for the sacred, and the blessing of the Church was a part of politics, nothing more. Indeed, in the late Renaissance in Italy and in the sixteenth century, the papacy itself often behaved in the manner of the political despots in defending its lands and its power.

Many of the men who led the entities, both large and small, had unusual personal qualities, even "genius." In this world, says Burckhardt, the exceptional person could succeed in politics, especially if he wanted some stability after attaining power. Burckhardt believed that Ludovico Sforza of Milan, "il Moro" (1452–1508), was the "most perfect type of the despot of that age." His means to gain power can only be regarded as immoral, for the modern state uses any means to achieve its ends, what is sometimes called "raison d'état." He knew the limits of power and cultivated diplomatic relationships with the Church and other important states. His court was the "most brilliant" on the continent, and cultivated another type of person "who, like himself, stood on their personal merits," artists, scholars, poets, and musicians. Leonardo stayed at his court. He founded an academy.

Along with Milan, Burckhardt cites Venice and Florence. Venice was both complicated and rational. It did much, including being (perhaps with Florence alongside) the birthplace of the use of statistics and data to help make civic decisions. Florence in the fifteenth century is given the honor of being fully "the first modern State in the world."

Not only were these states governed in a secular and rational manner, but they developed relationships between one another and with states outside Italy which made diplomatic relations also "a work of art." Italian foreign policy initiated the system of law and practice between states adopted by modern Europe. Military affairs were also systematized and thought about. These Italians, claimed Burckhardt, thus became the teachers of Europe.

Burckhardt was an admirer of Machiavelli, especially the Machiavelli of the *Discoursi* (1517), in which the Florentine studies Roman republican history as a possible model for his times: "of all who thought it possible to construct a State, the greatest beyond all comparison was Machiavelli."

In the modern era, Machiavelli is considered the first of the modern political thinkers. He is best known for the short work *The Prince* (1513; first published in 1532), in which he deals with the world of the despots. Machiavelli asked the question taken up by Burckhardt: how does a despot, a prince, obtain and retain power? His response is one which has given the term Machiavellian a bad name. However, Machiavelli's work is a moral one, for he speaks about what one needs to do in the name of state-building in states which are things-in-themselves, what price one pays for power.

His prince must calculate, sometimes lie, and sometimes use violence in order to keep power and order. In short, the state, which has its own reasons, has the need sometimes to kill and often to manipulate to further its interests. The modern state has a monopoly on violence, on incarceration, and on punishment.

The prince does evil, but in the name of something higher, the state itself. To be a prince is to enter into a very complex moral universe, to need to be feared rather than loved, to demand loyalty even though you sometimes must hurt some people, to use what Machiavelli calls necessary cruelty, rather than cruelty for its own sake. To serve the prince, one has only to recall the 1604 observation of Sir Henry Wotten (1568–1639), a diplomat and politician: "An ambassador is an honest man sent to lie abroad for the good of his country."

In the modern world, the autonomous state makes demands that are often exorbitant, including asking people to possibly die on its behalf. Robert M. Adams summed up Machiavelli's ideas regarding the prince and the state: "Do good when you can, do evil when you must; do both unhesitatingly, and don't lie to yourself about which is which."[42]

As Machiavelli stated:

> It is good to appear merciful, truthful, humane, sincere, and religious; it is good to be so in reality. [But] to preserve the state [the prince] often has to do things against his word, against charity, against humanity, against religion.[43]

When the state is a work of art, as we moderns have learned, sometimes it lacks moral restraint.

In these states, said Burckhardt, one can find the context for the development of the Italian person. And then the famous claim at the start of the second section, titled "The Development of the Individual": "To this it is due that he was the firstborn among the sons of modern Europe."

Burckhardt claims that the consciousness of people in the Middle Ages was qualitatively different from that of the Italian Renaissance:

> Man was conscious of himself only as a member of a race, people, party, family, or corporation—only through some general category. In Italy this veil first melted into air; an *objective* treatment and consideration of the State and of all things of this world became possible. The *subjective* side at the same time asserted itself with corresponding emphasis; man became a spiritual *individual*, and recognized himself as such."

He cites Dante as among the examples of the new man and new consciousness. The political ambience of Italy permitted this to happen and, in fact, encouraged it. In Italy in the fourteenth century there appear humans who are not troubled by their own unique identity, of being different from their neighbors: "Italy began to swarm with individuality."

The outcome of all this was the appearance of a private person, in some cases totally indifferent to community and politics. Wealth created enough of a leisure class of people who had the opportunity to cultivate individual thought. This occurred both among private citizens and those engaged deeply in politics. Individualism is associated with a new cosmopolitanism, a sense that the world

---

42 Robert M. Adams, "The Interior Prince, or Machiavelli Mythologized," in Niccolo Machiavelli, *The Prince*, trans. and ed. Robert M. Adams (New York: W. W. Norton, 243).
43 Ibid, 50–51.

is far larger than one's locality. Burckhardt quotes Ghiberti: "Only he who has learned everything is nowhere a stranger."

By the time of the fifteenth century, there is what Burckhardt calls "the many-sided man," what we now call the "Renaissance individual." There were statesmen and entrepreneurs who learned the classics, humanists who studied languages and the classics who also were learned in nature and science, painters who were well-read. All knowledge was the province of these people who delved deeply into both philosophy and life.

The example cited by Burckhardt is Leon Battista Alberti, whose varied life and studies included, among other matters, riding, speaking, music, civil and canon law, physics, mathematics, painting, Latin prose, and architecture, all done comprehensively and excellently. If Alberti was the first great model, he was succeeded by someone even more outstanding, Leonardo.

This was also both the result and the furtherance of a social equality in Italy which was different in kind from the class structure of the rest of Europe. Excellence was what it was about.

And what was sought? asked Burckhardt. Glory. Not necessarily riches, though that is fine; not necessarily power, though that is always a temptation. Poets vied to be the best. Scholars outdid one another. Artists and sculptors wanted to shine. In a world which was supposed to abandon temporal goals, these men of the Renaissance now sought fame and even celebrity.

A new cult arose in the adoration of the birthplaces and graves of famous men, not unlike that of those of saints in earlier times and still in much of Europe. Towns now began honoring ancient poets and philosophers who had resided there. The lives of eminent men of contemporary Italy were written up as examples of virtue. Many a despot, scholar, and poet made "preparations outwardly to win and secure fame," as ambition and the quest for greatness was part of the new social milieu. There was "a burning desire to achieve something great and memorable," which sometimes resulted in undertakings which produced more evil than good.

What is most important about Burckhardt's ideas about the new personality is the discovery of the self and the consciousness that goes with it. It is in this Renaissance that Burckhardt places the origins of the modern personality in all its variety and ambiguity. These Renaissance men are to be seen as the true ancestors of the people of 1859—self-creative, open, driven, experimental, troubled, and wanting to find excellence as they shaped their own lives. Burckhardt recognizes them as people who he knows. If some have called Baudelaire "the first modern," Burckhardt tells us that he has ancestors in the men of Renaissance Italy. The "I" now existed, a prelude to all that came afterward.

Only now, in the third part, does Burckhardt get to "The Revival of Antiquity." He undercuts nearly all that had been written earlier about this matter, putting the revival of the classics at the center of the times. Yes, it is through the revival of the classics that the new spirit occurs. However, it was important but not necessary: "the essence of the phenomena might still have been the same without the classical revival." Hence, "it was not the revival of antiquity alone, but its union with the genius of the Italian people, which achieved the conquest of the western world."

The Church had been the institution and Christianity the idea that had for centuries been at the center of Europe. Beside this, claims Burckhardt, there arose "a new spiritual influence which... became the breath of life for all the more instructed minds of Europe." Burckhardt emphasizes throughout the spiritual quality of the ideals of the men of the Renaissance in Italy. Many have said that the Renaissance is still part of the Age of Religion. Perhaps. But for Burckhardt, a modern who well knew the ideas of both the rationalists and the romantics, religion does not have a monopoly on spirituality. The new humanist culture vies with the Church and, eventually, contributes to new secular ideals.

Hence, the classical revival is only a piece of the Renaissance and not its cause. Before that, he states "a development of civic life was required," and this first occurred only in the Italian peninsula. Moreover, a new class of person appears as well, the burgher, the city dweller engaged in commerce, which is now a power alongside the nobles. It was necessary that these two groups, nobles and burghers,

> should first learn to dwell together on equal terms, and that a social world should arise which felt the want of culture, and had the leisure and the means to obtain it. But culture, as soon as it freed itself from the fantastic bonds of the Middle Ages, could not at once and without help find its way to the understanding of the physical and intellectual world. It needed a guide, and found one in the ancient civilization, with its wealth of truth and knowledge in every spiritual interest... The spirit of the people, now awakened to self-consciousness, sought for some new and stable ideal on which to rest.

Burckhardt did not elaborate on the matters of class and economics as much as would be done by scholars of the twentieth and twenty-first centuries, but here is the first reference to that group of people who would, in 1859, be recognized as leaders in modernity, the new bourgeoisie, those who would be celebrated by Manet and his fellow Impressionists, Baudelaire and modernists, even praised (though they would be wiped out, they claimed) by Marx and Engels, and given agency by Mill. And if Europeans in the mid-nineteenth century thought themselves at the head of progress, Burckhardt noted that "armed afresh with [the classical] culture, the Italian soon felt himself in truth citizen of the most advanced nation in the world."

Rome became the model and the ruins of Italy were a kind of classroom for scholars and writers. Now, even before the printing press, the new knowledge, both from the classics and contemporary writers, was diffused as far as possible. Book-finders were significant, and libraries were built as the acquirement of books became a celebrated activity. Textual criticism became important as languages were acquired and used for scholarship. Universities began to grow in the thirteenth and fourteenth centuries.

Now the group of people called humanists appeared. They were the mediators between antiquity and modernity at the time, and they challenged the authority of the Church. Italian glory was now not in the Church but in antiquity. These new men became the teachers of the families of the powerful despots. Heretofore, works on how a prince should be educated and behave were written by figures in the Church; now it is the humanists who undertook that task.

Humanists often became part of the civic structure of states as they helped with the correspondence of the state and in the rhetoric of officials on public occasions. It was humanists who now wrote history, composed dialogues, and became poets. As far as Burckhardt was concerned, by the fifteenth century, "the Middle Ages were over for Italy." However, he does record what occurred in the first half of the sixteenth century, when the humanists and their culture were overcome by the events and leaders of the Counter-Reformation.

Burckhardt next writes about "The Discovery of the World and of Man." He does not cite Michelet (the entire work, as scholarly and knowledgeable as it is, has no footnotes), but it is highly likely this fourth section's title is taken from him. Again, Burckhardt wants us to remember that the new state and the new personality are at the core of the era.

The new geographical knowledge and the natural sciences are acknowledged. Burckhardt, however, emphasizes the interest in nature in a variety of ways. In Italy there began the cultivation of botanical gardens and the interest in having a variety of plants and species. This was complemented by collections of a variety of foreign animals as a way of studying nature. Having live lions was popular in several states. Larger menageries included many "exotic" animals, including horses and dogs. The mid-nineteenth century was a time of great advances in the study of nature, and Burckhardt theorizes that it was in Renaissance Italy that "the foundations of a scientific zoology and botany were laid."

The key for Burckhardt was that the Italians were drawn to nature as "the first among modern peoples by whom the outward world was seen and felt as something beautiful." Artists and poets desired to represent nature, again something deeply felt in the mid-nineteenth century. Most importantly, Burckhardt remarks, nature no longer had any association with anything sinful. As he often does, Burckhardt cites Dante, for him the first modern. Dante and Petrarch saw nature as part of what Burckhardt calls "the human spirit." They praise the air and the light, they admire its grandeur, and they, like the Romantics after them, climb mountains as a way of participating in the enjoyment of natural beauty. Burckhardt remarks on the poetry of Italy, which "is rich in proofs of the powerful effect of nature on the human mind."

The discovery of man is even more significant. There was an effort to observe and describe humans and human nature in all of its depth and variety. Even more, there was a great leap which affects historical periodization. Dante's exploration of humanity is for Burckhardt "the boundary between medievalism and modern times." It is in the ideal of the individual and the consciousness of our own inner life that makes Italy unique and makes the Renaissance a period in its own right.

Not only new content, but new forms are important. In poetry it is the invention and the new forms of the sonnet which gave special qualities to Italian thought, by giving poets a tight means to lyrically express thoughts and emotions in a clear manner. Yet again, Burckhardt is implicitly engaging in a dialogue between past and present, for at the time he was writing Baudelaire and others were developing their own new poetic and prose forms to express their understanding of the human experience of modernity.

New genres are developed. There was an attempt to deal with humans in history in an accurate manner, says Burckhardt, "according to his inward and outward characteristics." Biographies were written, especially of unusual men of the times, in order to learn about the characteristics of greatness, much as in ancient times. In the medieval period, stress was placed on the lives of the saints. In the Renaissance, it was poets, scholars, and statesmen. And autobiographies were written, sometimes dealing with family histories, sometimes dealing with a single individual, fully presenting a personality. Burckhardt does not mention it, but of course it is in the Renaissance that the portrait of secular individuals was initiated.

In describing human beings, collectively in cities and states, and individually, there is a further breakthrough made in the attempt to find and represent that which is beautiful. A new aesthetic is attempted by such writers as Boccaccio and, in the sixteenth century, by the important work of Firenzuola on the nature of female beauty.

Daily life became a subject of interest and of poetry and art—for example, hunting, journeys, ceremonies, and country life. Burckhardt argues that it is only in Italy at the time that the peasant was understood to have some personal freedom, dignity, and the right to settle where he chose. Hence, all classes became the subject of literature and art.

Burckhardt closes this section by stating clearly his main thesis: it was in Italy at that time "that men and mankind were... first thoroughly and profoundly understood. This one single result of the Renaissance is enough to fill us with everlasting thankfulness. The logical notion of humanity was old enough—but here the notion became the fact." And to provide a coda, he summarizes and quotes a part of the great "Oration on the Dignity of Man" by Pico della Mirandola. "To thee alone," Pico has the Creator say to Adam, "is given a growth and development depending on thy own free will. Thou bearest in thee the germs of a universal life." We are now far closer to Milton than we are to Aquinas.

Burckhardt then went on to discuss "Society and Festivals," giving attention to that which was mainly ignored by the historical establishment.

The key to this was something that followed from the new state and new concept of personality. Now,

> Italian customs at the time of the Renaissance offer in these respects the sharpest contrast to medievalism... Social intercourse in its highest and most perfect form now ignored all distinctions of caste, and was based simply on the existence of an educated class as we now understand the word. Birth and origin were without influence, unless combined with leisure and inherited wealth.

Nobles and the new bourgeoisie lived together inside the walls of the cities.[44] With the influence of the new humanism, personal merit mattered more than birth, very much unlike the north.

The result was something that Burckhardt identifies as "modern," a society based on culture and wealth, the wealth to be used in service of the culture. Hence, the individual, now that he appeared, needed "to make the most of his personal qualities, and society to find its worth and charm in itself." There is yet another work of art to be found in the Italian Renaissance: "the demeanour of individuals, and all the higher forms of social intercourse." Humans had the opportunity to shape themselves, an extremely modern concept.

Matters only discussed by professional historians in the last several decades are now taken up by the mind of Burckhardt. Fashion becomes important—dress, one's outward appearance, and the use of various techniques to make one's skin and hair lovely. Perfumes, says Burckhardt, "went beyond all reason-

---

44 The Italian sovereignties at the time, Florence and Milan for example, included both the walled cities and the hinterland.

able limits... At festivals even the mules were treated with scents and ointments..."

As well, Burckhardt discusses a matter dear to those reshaping cities and aspiring to modernity in the mid-nineteenth century, the idea of the importance of cleanliness, which he recognizes as "indispensable to our modern notion of social perfection, which was developed in Italy earlier than elsewhere." He notes that the evidence for this doesn't have the kind of clarity that a political or diplomatic document might have. Yet, it is clear from the literature and evidence of the time that cleanliness was valued and practiced, especially at meals. He also notes that "'German' was the synonym in Italy for all that is filthy." The north was viewed as coarse.

Language is discussed as the various Italian dialects developed in a dynamic, changing society. Purists could not manage to keep the language fixed. His key point is that great respect and esteem were given to those who wrote and/or spoke well and with feeling. Only later, after the end of the Renaissance, could the Accademia make Italian seem "like a dead language."

Manuals of courtesy were now written to let people know what was proper etiquette. As well, there came to be the courtier, best described by Castiglione, "the ideal man of society," whose relationship to the prince is moral and independent. He is guided not by fealty, but by honor. He does not pander to the court, the court exists for him. He cultivates sensibility, ideal love, and excellence in all things. Festivals, including their processions, became more secular as time wore on. If anything they looked to the classical world for their costumes and themes.

Few commentators on Burckhardt note that he included in his work, a cultural history as he and others called it, a good deal of guidance to the cultivation of social history, a century before it became an ordinary part of historical study. Not only was his subject different from other academic historians. His methodology demanded a new use of the historical imagination.

The last section, "Morality and Religion," is one which Burckhardt introduces with a hesitant prologue. He is uncomfortable judging others, and he follows the German Romantics in believing each group of people has its own identity and beliefs. Hence, he states, these are notes about the period in Italy, about their morals and religious beliefs, rather than opinions. As he asks of the reader, "What eye can pierce the depths in which the character and fate of nations are determined?"

Still, Burckhardt is disturbed by the morality of later Renaissance Italy. If he begins his understanding of the Renaissance with the divine Dante, it ends with the realist Machiavelli. He quotes the latter: "we Italians are irreligious and corrupt above all others... because the Church and her representatives set us the worst example."

There is now, at the start of the sixteenth century, "a grave moral crisis." This is because all restraints seem lost in a world which only has a sense of honor as its basis. Burckhardt even states that this matter is something which is part of his own day. For holiness, once the works of antiquity were absorbed, the Italians substituted "the cult of historical greatness," hence they considered some matters which were vices to be unimportant.

He makes a comparison, the only one in this work, with the Enlightenment. The implication is that the emphasis on individuality and honor, while it produced remarkable works and events, also had the capacity to foment vice and evil. Indeed, he takes the time to fully discuss destruction and vengeance. The Italian, whose own vice was "excessive individualism," ended up egoistically only taking his own interest into account in social and political affairs.

How to explain it? Burckhardt used a category beloved by Marx, historical necessity:

> But this individual development did not come upon him through any fault of his own, but rather through an historical necessity. It did not come upon him alone, but also, and chiefly, by means of Italian culture, upon the other nations of Europe, and has constituted since then the higher atmosphere which they breathe. In itself it is neither good nor bad, but necessary; within it has grown up a modern standard of good and evil—a sense of moral responsibility—which is essentially different from that which was familiar to the middle Ages.

Hence the Italian of the Renaissance is also the first modern to confront the ending of a *Weltanschauung* and the openness it provides.

As for religion, Machiavelli provides the response. There was contempt for the hierarchy and for the institution of the Church. Burckhardt provides enough evidence for the reader to come to the conclusion that in the Italy of the time the Church emulated the new humanists and despots rather than the other way around. The belief in God wavered. The need for salvation receded as people lived in this world and in the present. Indeed, the response to religion and the growth of individualism set the stage for later developments in Christianity. One very positive effect of the Italian Renaissance was a tolerance and understanding of Islam different from the rest of Europe.

Burckhardt concludes that humanism was fundamentally pagan. The basic dogmas of Christianity dissolved under the onslaught of ancient knowledge and the agenda of poets, artists, statesmen. These new men, laments Burckhardt, "repent of nothing":

> The passive and contemplative form of Christianity, with its constant reference to a higher world beyond the grave, could no longer control these men. Machiavelli ventured still further, and maintained that it could not be serviceable to the State and to the maintenance of public freedom.

Theism or deism, whatever it is called, first appeared then. Deism, the idea of a first mover who set the planet in motion and then withdrew, did away with Christian fundamentals in religion. Theism, simply the belief in one or more gods, was also present. It could include Christian ideas, but it could also be understood without them. Both are modern concepts, arising again in the eighteenth and nineteenth centuries in western Europe. Both were prominent in the world of 1859.

The Renaissance we study is the creation of Burckhardt. He sought to find what his own age would find, as he put it, "worthy of study." He found the origins of modernity in Italy in the fourteenth and fifteenth centuries and he knew that this development was central to understanding what Europe was in the middle of the nineteenth century. It may have been effaced for a time, but it again came into being after the French and industrial revolutions.

Burckhardt stayed at the University of Basel for the remainder of his long life. It was a small university, not major at all, but he was content to live in Switzerland and to be something of an outlier, though immensely respected, in the historical profession.

In 1872, Burckhardt was offered Ranke's chair at Berlin after Ranke's retirement, the most prestigious chair of history in Europe. He turned it down. Burckhardt still had differences with Ranke and his followers, most especially in understanding the relationship between culture and the state. Though conservative, Burckhardt did not believe that state power necessarily enhanced culture or was a sign of greatness. The new German state did not appeal.

Another important figure appeared at Basel in 1869, a young new professor of classics, Friedrich Nietzsche. He and Burckhardt struck up a friendship at the time, and Burckhardt influenced Nietzsche's first major study, *The Birth of Tragedy* (1872). When Nietzsche later had occasion to reflect on the uses of history, he too talked about finding a history that mattered to the present. The friendship, which went on for some time and includes interesting correspondence between the two, eventually dissolved as Burckhardt was repelled by some of the later pronouncements of Nietzsche.

Burckhardt's contribution to the intellectual world of his time was profound and important. By defining a new period and a new type of human being, he helped his contemporaries to understand the world of 1859, its accomplishments, its issues, and its paradoxes.

Chapter V
John Stuart Mill: Liberty and Modernity

The terms liberalism and conservatism were part of the political discourse of mid-nineteenth century Europe, along with socialism, communism, empire, and authoritarianism. However, the two ideas were imprecise designations, since they have changed their core meanings depending upon the time and the context.

Still, liberalism and conservatism signalled a posture, if not fully coherent philosophies. At one end of Europe, Britain in the west, to be liberal meant to be open to change, to support laissez-faire economics in theory if not sometimes in practice, to follow utilitarian concepts and to argue for the autonomy of the individual. To be conservative was to align with many of the ideas of Edmund Burke, including the importance of traditional institutions such as the monarchy, parliament, and the established church, to believe in order and be fearful of rapid change, to emphasize the coherence of society and slow development, and to challenge the ideas of natural rights and constitutionally limited government.

At the other end of Europe, in Russia and elsewhere, conservatism was seen as a bulwark against disorder and revolution, a doctrine which elevated custom and tradition far ahead of the new and the need to change. Liberals were seen as dangerous, for they would transform what the conservatives regarded as a natural order which had developed over time into a world which uprooted all they believed was given and civilized.

If Burke was the prophet of British conservatism, it was Metternich who was the defender of the order and the peace of the eastern part of the continent. And, indeed, the major settlement of the first half of the nineteenth century, one which brought an order and some peace after the era of wars from 1792–1815, the treaty made at the Congress of Vienna, was one which ratified a conservative Europe against the challenges stemming from the French Revolution. The revolutions and upheavals of 1848 only substantiated this belief for conservatives.

In 1859, liberalism as an idea was not only challenged from the right. To the left were various socialists, anarchists, and communists, all finding liberalism too bourgeois for their tastes and beliefs. Nonetheless, liberalism had many supporters, from those like Mazzini who incorporated many liberal ideas into his version of nationalism to others who liked its constitutional base and direction of peaceful change.

Early that year, what would become one of the great works of the liberal tradition was published, an essay of roughly 125 pages by the British philosopher John Stuart Mill, called *On Liberty*.

In his *Autobiography* (1873), Mill stated:

> The 'Liberty' is likely to survive longer than anything else that I have written (with the possible exception of the 'Logic')... [It is] a kind of philosophic text-book of a single truth, which the changes progressively taking place in modern society tend to bring out in stronger relief: the importance, to man and society, of a large variety of types of character, and of giving full freedom to human nature to expand itself in innumerable and conflicting directions.[45]

Mill was by 1859 one of the most prominent philosophers in Europe, having edited and published a number of major works, including *A System of Logic* (1843) and *The Principles of Political Economy* (1848).

What is also notable about Mill, and is one of the guides to understanding the philosophical principles of *On Liberty*, is his own background and education, as told in his *Autobiography* and by others involved.

Mill was, by every account, a rarity as a child in that he was educated into and became one of the few recorded philosophical prodigies in the West, perhaps the only one of his kind. In a field where one usually matures late in life, far later than mathematicians, musicians, and physicists, Mill became recognized as an unusual intellect and he published interesting philosophical works by his late teens and early twenties.

He was of utilitarian stock, a background that he critiqued and revised. Yet it continued to influence him all of his life. His father, James Mill, was the second best known utilitarian thinker in Britain in the early nineteenth century, next to Jeremy Bentham. The two men were close friends.

James Mill took his eldest child's education in hand. The genius was homeschooled. It was a rigid process by a father who was relentlessly rationalist.

John started learning Greek at three and Latin at eight, by which time he was reading Herodotus and Xenophon in Greek, and was responsible for teaching the language to one of his younger siblings. By the age of 12 he had read more of the classics than most university students. In mathematics, he had at age 12 learned "elementary geometry and algebra thoroughly, the differential calculus and other portions of the higher mathematics far from thoroughly."[46] He liked history and read much on his own. He also began the formal study of logic at that time.

---

45 *Essential Works of John Stuart Mill*, ed. Max Lerner (New York: Bantam Books, 1961), 148–49.
46 Ibid., 17.

James Mill's *History of British India* was published in early 1818. His eleven-year-old son helped with the galleys.

In Britain, the division of knowledge was in two parts, rather than by separate disciplines. The sciences and mathematics were called natural philosophy, and what we now call the humanities and social sciences were grouped under the heading of moral philosophy. Moreover, to this day there is in Britain the study of what is called political economy, which John began studying in 1819, including the work of a family friend, David Ricardo, with whom he discussed Ricardo's writings.

Another close family friend was Jeremy Bentham, the leading utilitarian and the inventor of the formula of "the greatest happiness of the greatest number." He, too, took an interest in John's mental development.

In May 1820 John was sent to live in France with the family of Samuel Bentham, Jeremy's brother. He continued his studies there. In addition, he recalled, he obtained a "familiar knowledge of the French language," and, more importantly, he learned about "the free and genial atmosphere of Continental life." When he returned from France to England he read Pierre Étienne Louis Dumont's French edition of Bentham's *Traité de Législation*, which he said was "an epoch in my life."

Mill became a member of the group of utilitarian thinkers, known as philosophical radicals, on his return from France in 1821. In the next few years he published in the *Westminster Review* regularly, as well as in the *Parliamentary History and Review* in its three years of existence from 1825–28. He founded a study group, the Utilitarian Society, and took part in public debates. In 1825–26 he took on the responsibility of editing Jeremy Bentham's papers on the matter of evidence, published in five volumes in 1827 as *Rationale of Judicial Evidence*.

Hence, by 1826, at the age of 20, Mill was one of the best educated people in Europe, and a person of some accomplishment. He had a goal after reading Bentham: "to be a reformer of the world."

There are two major British novels which tell what might happen to someone raised as was Mill. Mary Shelley's Dr. Frankenstein in her 1818 novel is a great natural philosopher deficient in moral understanding of the consequences of his acts. He suffers from melancholia and he makes choices in his life which cause great damage to himself and his family.

The other novel is Dickens' *Hard Times* (1854), in which Dickens invents a father, Thomas Gradgrind, who is a parody of James Mill and other rationalists. He only knows about facts, is wholly uninterested and deficient in feelings, and his two children experience great mental and emotional hardships as a result of his kind of education.

John Stuart Mill suffered what he termed a "mental crisis" in the fall of 1826, at the age of 20. As if ambushed, he felt he was under a cloud, that life was not at all satisfying and there were no remedies. He asked himself if his program of reform were put in place, would that be satisfying, would that lift the gloom of his mood, and "an irrepressible self-consciousness distinctly answered, 'No!'"[47]

Mill understood that he should obtain pleasure in reform. But he could not shake the feeling that his life now lacked meaning, depth, and purpose. The depression lasted six months, into the spring of 1827. Mill relates that he went about his responsibilities and relationships in a mechanical manner, as a kind of automaton, "by the mere force of habit."

In retrospect Mill blamed the episode on the nature of his education and upbringing, his having been raised by a father who only valued logic and reason and displayed no real feelings. Not only was James Mill thoroughly bored with his wife, but John's mother seemed to be absent in his emotional development. Indeed, even in the *Autobiography* his mother is not present. In telling of his birth, Mill stated: "I was born in London, on the 20th of May, 1806, the eldest son of James Mill...." (in Dickens' *Hard Times*, Mrs. Gradgrind has no first name, and we are told that Thomas Gradgrind married her because she lacked "fancy"). And Mill reflected, in a famous sentence, "If I had loved anyone sufficiently to make confiding my griefs a necessity, I should not have been in the condition I was."

He did what Mill did his whole life. He read. In this case it was Jean-François Marmontel's *Mémoires*. The passage in which Marmontel describes his father's death when he was a boy and his determination to help his grieving family moved Mill to tears. He happily realized that he had not become someone totally unable to feel emotion. He then began to find some joy in simple pleasures—in the sun and the sky, in conversation, in being involved in public affairs—and he now again could find "excitement, though of a moderate kind."

Mill then listened to music again, in which he found pleasure. The cloud fully lifted with his discovery of the poetry of Wordsworth and in taking delight in the beauty of nature:

> What made Wordsworth's poems a medicine for my state of mind, was that they expressed, not mere outward beauty, but states of feeling, and of thought coloured by feeling, under the excitement of beauty. In them I seemed to draw from a source of inward joy.

Mill now recognized he could engage in that Victorian endeavor, a course of self-improvement, whereby he could become a fuller and deeper human being. In

---

47 Ibid., 83.

this moment, he realized that the rationalist program had its limits. He never regretted his education, but he knew it had ignored a deep part of human nature.

Mill gained much from the crisis. Afterwards, like some who come out of a deep depression, he believed himself to be a better and more complete person. He gained a deep empathy for others sometimes ignored by the counting utilitarians and their "greatest happiness for the greatest number." And he resolved to make his program of reform include the right of individuals to find happiness in their own manner.

Mill himself stated that he believed two important matters resulted from the episode. He learned that, paradoxically, happiness for oneself can only be achieved by pursuing a course of action which helps others. We are, as Aristotle said, social animals, not simply a collection of individuals. Second, he believed that that he, "for the first time, gave its proper place, among the prime necessities of human well-being, to the internal culture of the individual... The cultivation of the feelings became one of the cardinal points in my ethical and philosophical creed."

Though he did not speculate on the matter, Mill's depression is part of something that is essential to understanding the consciousness of modernity. After all, Baudelaire had his periods of unhappiness, Dostoyevsky's characters are sometimes mad, Turgenev's superfluous men are deeply sad, Darwin had his lethargy, and many mid-nineteenth century novels and plays deal with eccentricity, madness, and insanity, not only among males but in the lives of oppressed females.

What was happening was partly a result of the secularization of life. Now there was a void to be filled, and those individuals who no longer relied on traditional religions or on a fixed ideology were left to their own devices. Among other matters, modernity is about consciousness and self-reflection. It is about acting and watching oneself act at the same time. And it is about to be always in a state of Becoming. Joy is to be found, but gloom lurks in its wake as well.

Much that Mill learned from his crisis is part of what went into the writing of *On Liberty*, begun in a short version in 1854 and published in 1859.[48]

---

[48] Mill did find the love of his life in Harriet Taylor, who became his wife in 1851. Their story is an unusual and fascinating one, and the interested reader can find it many works. Mill regarded Taylor as a collaborator in his intellectual life and development. Sadly, Taylor died in November, 1858, just before the publication of *On Liberty*. Mill dedicated the book to her: "To the beloved and deplored memory of her who was the inspirer, and in part the author, of all that is best in my writings—the friend and wife whose exalted sense of truth and right was my strongest incitement, and whose approbation was my chief reward—I dedicate this volume. Like all that I have written for many years, it belongs as much to her as to me; but the work as it stands

The opening pages of the essay make clear Mill's subject and introduce important ideas. He is not discussing the matter of free will and necessity. Rather, he is interested in what he calls "civil, or social liberty," that is, "the nature and limits of the power which can be legitimately exercised by society over the individual." It is important to keep in mind that Mill is not just asking about the relationship between the individual and the political authority. He is a liberal thinker who, like conservatives, recognizes the importance of both society and government. It is also important for readers and commentators to note that Mill is not writing a treatise on liberal theory or the definitive work on liberalism. Many have suggested Mill should have written a different essay, and then challenge him for not doing so.

In more concise terms, Mill asks about "the struggle between liberty and authority." In this formulation he discusses stages of development of the tension or struggle, giving a brief history of arrangements from Rome to the present which regulate freedom and power, for the issue of the balance between liberty and authority is a constitutional one, spelled out in the customs and laws of various sovereignties.

Mill makes clear throughout the essay that his concerns in 1859 are several features of modernity: the rise of the masses and the nature of liberal democracy; the rise of the power of state authority and how it relates to individual liberty; and the importance to society and polity of the freedom of the individual. In this, Mill is ahead of his time, for these matters would become central decades later with the rise of fascism, bolshevism, and authoritarianism.

As well, Mill in the epigraph of the essay tells us that we are in a new era of thinking about human nature. The sentence is taken from Humboldt's *Sphere and Duties of Government* (1854): "The grand, leading principle, towards which every argument unfolded in these pages directly converges, is the absolute and essential importance of human development in its richest diversity."

Like Baudelaire and others, Mill now sees humans as what he terms "progressive beings." We have a nature, but it is not fixed. Individuals develop and change as they mature and grow. Hence, Mill is especially interested in protecting the ability to do so in any society or state. We cannot fully express our

---

has had, in a very insufficient degree, the inestimable advantage of her revision; some of the most important portions having been reserved for a more careful re-examination, which they are now never destined to receive. Were I but capable of interpreting to the world one-half the great thoughts and noble feelings which are buried in her grave, I should be the medium of a greater benefit to it than is ever likely to arise from anything that I can write, unprompted and unassisted by her all but unrivalled wisdom."

diverse humanity unless we have the option to shape our lives as we see fit, as long as we do no harm to others.

The implications of this idea can be revolutionary, depending upon the context. For example, Mill argued and wrote on behalf of the emancipation and equality of women, believing that this would be useful to all of us as well as to individuals. And, of course, this idea has implications for colonialism and imperialism, anticipating in England the twentieth century arguments of people like Woolf and Orwell.

For his own time, Mill identified a development in democracy which could turn it into a tyranny, the "tyranny of the majority." The concept he acknowledged was taken from his reading of Tocqueville and he found it of vital importance. The will of the majority, he argues, cannot be used to tyrannize a minority, as a means of oppression. In a political sense, this means that Mill accepts the idea that political authority must be limited. Governments do not possess the whole of political power and then return part of it to their members, as theorized by Hobbes.

One of the many issues raised by commentators of *On Liberty* is the extent to which Mill kept to the tradition of utilitarianism in his arguments and assumptions. In this case, and in others, Mill is not in favor of the greatest happiness of the greatest number. Rather, he will move to a tacit acceptance of the ideas of natural rights and contract theory in determining the limits of authority. He prefers the French-American tradition to that of Britain. Individuals retain some rights apart from the powers given to the state. These rights cannot be trampled upon, whatever the majority asserts. This is one of the ways to protect individuality and diversity from the conforming nature of authority.

Mill carries the tyranny of the majority one step further. It is not only political, it can be social. Society can make demands on individual behavior and punish those who do not conform. Indeed, Mill believes that social tyranny can be "more formidable than many kinds of political oppression… penetrating much more deeply in to the details of life, and enslaving the soul itself." Social despotism can be as bad as political despotism. Non-conformity must be tolerated and left alone. Mill fears that the trend in modernity might be one which smashes and punishes individuality.

This clarifying opening also makes reference to another concept important to Mill's argument, that of "self-regarding acts." Mill asks that any act which affects only the actor be permitted and protected. He suggests that some parts of our behavior are ours alone—for example, what we read, when we sleep, where a *flâneur* will walk, who we might write letters to, etc.

The opposite of self-regarding acts is those which are other-regarding, which affect others. For example, how we behave towards others, what social ethic we adopt, whether we live up to our obligations towards others in our lives and in the community.

The issue of self-regarding acts is one which has been argued ever since. There are those on the left and the right who will suggest that no act is self-regarding—for example, if you wear a garment made by someone who is oppressed in a factory, that is a political matter; if you eat meat you accept the killing of animals, etc. Others say that some acts remain self-regarding in the main, if not fully so, and should be protected. And there are those who claim that some kinds of behavior may be self-regarding and others other-regarding, but there is a grey area in the middle which is difficult to define outside of a specific context.

The idea of self-regarding acts needs context if we are to deal with it carefully. Reading certain books in Canada is a matter of taste. Reading those same books in an authoritarian state—Saudi Arabia, Turkey, Iran, for example—is a political act. Wearing certain clothes in some places is self-regarding; in others it is other-regarding.

Generally, the issue of self-regarding acts has been defended by those interested in claiming the rights of the individual. Whether I today write part of a chapter on Mill or read a novel is my choice. Whether I decide on a career in business, carpentry, law, plumbing or banking is my choice. I retain the ability to decide how I might live my life as a progressive being. The key for Mill is to err on the side of individuality, even eccentricity, for reasons related to both natural rights and utilitarianism. As he put it, "the free development of individuality is one of the leading essentials of well-being." He is among the first to see the growth of conformity in capitalist and industrial society and to argue against it.

After finishing with the subject of the essay, Mill then turns to its object. He asserts "one very simple principle," not simple at all, as the object of his work: "That principle is that the sole end for which mankind are warranted, individually or collectively, in interfering with the liberty of action of any of their number is self-protection." No one, in a civilized society, may be permitted to harm others. However, you cannot compel someone to do something because you believe it will do him good. You can argue with him, you can plead, you cannot force: "In the part [of his conduct] which merely concerns himself, his independence is, of right, absolute. Over himself, over his own body and mind, the individual is sovereign."

Mill then goes on to define what he calls "the appropriate region of human liberty," worth quoting because it has become something cited by all defenders of human rights ever since:

This, then, is the appropriate region of human liberty. It comprises, first, the inward domain of consciousness; demanding liberty of conscience, in the most comprehensive sense; liberty of thought and feeling; absolute freedom of opinion and sentiment on all subjects, practical or speculative, scientific, moral, or theological. The liberty of expressing and publishing opinions may seem to fall under a different principle, since it belongs to that part of the conduct of an individual which concerns other people; but, being almost of as much importance as the liberty of thought itself, and resting in great part on the same reasons, is practically inseparable from it. Secondly, the principle requires liberty of tastes and pursuits; of framing the plan of our life to suit our own character; of doing as we like, subject to such consequences as may follow: without impediment from our fellow-creatures, so long as what we do does not harm them, even though they should think our conduct foolish, perverse, or wrong. Thirdly, from this liberty of each individual, follows the liberty, within the same limits, of combination among individuals; freedom to unite, for any purpose not involving harm to others: the persons combining being supposed to be of full age, and not forced or deceived.

He states that no society is free, whatever its form of government, unless these liberties are defended. All this is not new, claims Mill, but these liberties are being challenged as never before in 1859 by societies, governments, and religious authorities. What would he have said if he had lived in our world of surveillance capitalism?

Mill then gives reasons why free speech is not only necessary for individual development, but is useful to society as a whole. In this section he moves back to a utilitarian argument, buttressing the natural rights assumptions of his basic ideas.

"First," he states, "the opinion which it is attempted to suppress by authority may possibly be true." However wise government authorities might think they are, however certain they are about the truth, there is no such thing as *"absolute* certainty." A condition of asserting that something is true is the ability to test it in the marketplace of ideas before you act on that truth. Here, Mill is asserting the importance of discourse, something part of both liberal and democratic theory. Without discourse, there is no way of knowing anything or improving one's ideas.

The utility of an opinion or idea itself is something also open to contest and requires discussion. No one idea or person or government can be assumed to be infallible. Mill challenges religious dogma in society, though he does not challenge the right hold whatever religious view one supports. Here he introduces a kind of character type that is important to modernity, the heretic, citing Socrates and Jesus, among others.

The rooting out of heresy, far from aiding progress, thwarts it. It is, for Mill, both intolerant and a violation of the rights of the heretic, and of the need for discourse to progress. Indeed, heresy is useful, he argues. Heretical ideas are sometimes suppressed truths, countering dogma.

The heretic as someone challenging authority is a common figure in modernity, often someone who is marginalized, oppressed, sometimes killed. Mill did not see the growth of the authoritarianism of the twentieth century. Yet his discussion of heretics recalls that of Yevgeny Zamiatin, the renegade Bolshevik who authored the dystopia *We* in 1921, a book not published in Russia until 1988.[49] Heretics are useful, said Zamiatin, who ended his life in exile. Some of his comments on the matter include:

> "Heretics are necessary to health; if there are no heretics, they should be invented."[50]
> "Heretics are the only (bitter) enemy against the entropy of human thought."[51]
> "Harmful literature is more useful than useful literature, for it is antientropic, it is a means of combating calcification, sclerosis, crust, moss, quiescence. It is utopian, absurd... It is right 150 years later."[52]

The heretic, the eccentric, the maverick, all are defended by Mill, who understands that all those designations are both political and social constructions. He knew that some of these people might have little to offer, but he also knew that some—the Baudelaires, the political dissenters, the Darwinian scientists, and others—were the basis of our greater understanding of humanity and society. "Genius can only breathe freely in an *atmosphere* of freedom," he wrote in *On Liberty*.

Mill himself had no religious beliefs. He was an unbeliever, rather than an atheist, for as he put it, he was a rare example of someone in Britain who had not "thrown off religious belief, but never had it." He studied religious philosophy and the history of religions as part of his background in moral philosophy. However, we should remember that in Britain in his time there were legal limits on those who were not members of the Anglican Church, still the established church of England.

---

49 It is of interest in this context to note that Mill is credited with inventing the term dystopia. Mill was elected to parliament as a Liberal in 1865 and served until 1868. In a speech of March 12, 1868, Mill responded to the Irish land policy of the conservative government: "It is perhaps too complimentary to call them Utopians, they ought rather to be called dys-topians or caco-topians. What is commonly called Utopian is something too good to be practicable; but what they appear to favour is too bad to be practicable."
50 "Literature, Revolution and Entropy," in *A Soviet Heretic, Essays by Yevgeny Zamyatin* (Chicago: University of Chicago Press, 1970), 109.
51 Ibid., 108.
52 Ibid., 109.

The oath taken by members of the British parliament excluded all but Anglicans for some time. In 1829 an oath acceptable to Roman Catholics was included in the Roman Catholic Relief Act. In 1858 new legislation permitted Protestants and Jews to serve as MPs. A single oath for members of all religious beliefs was introduced in 1866, and in 1888 permission was given to those objecting to the taking of an oath on religious grounds to make another affirmation.

In 1871 the Universities Test Act was passed which abolished religious "tests," thus permitting Roman Catholics, non-conformists, and non-Christians to take on the roles of professors, fellows, and students in the Universities of Cambridge, Oxford, and Durham.

Another great Liberal of the next generation, Lord Acton (1834–1902), was an Anglo-Catholic. In 1895, he was appointed to the highly prestigious chair of Regius Professor of Modern History at Cambridge. As was the custom, he gave a public Inaugural Lecture. The opening of the lecture reminded his audience of an irony around his appointment:

> Fellow Students—I look back to-day to a time before the middle of the century, when I was reading at Edinburgh and fervently wishing to come to this University. At three colleges I applied for admission, and, as things then were, I was refused by all. Here, from the first, I vainly fixed my hopes, and here, in a happier hour, after five-and-forty years, they are at last fulfilled.[53]

Two or more conflicting opinions, Mill also argued, often share the truth between them. Through discourse you might arrive at a higher truth. Indeed, several opinions might be necessary in any society. The example he gave was the importance of having diverse political views. Having both a party of order and a party of reform makes for a healthier polity. Politics is dialectic, not dogma.

The claims of religion were countered by Mill by the claims of tolerance, rights, and utility. That a group regards certain doctrines to be sacred does not give them any higher status than other person's beliefs. He does note that he has seen that Christians do not behave in a manner consistent with their code. Examples he gives include that the poor are blessed, and that it is "easier for a camel to pass through the eye of a needle than for a rich man to enter the kingdom of heaven." Christians do believe their doctrines, he agrees. However, the great majority of Christians do not act on them.

Even if we accept the validity of many parts of the Christian ethic, Mill argues that it is an incomplete code. Hence, it must be supplemented by other be-

---

[53] Acton's Inaugural Lecture on History | Online Library of Liberty, https://oll.libertyfund.org › page › acton-s-inaugural-lecture.

liefs and revised accordingly. Like other doctrines Christianity has only a part of the truth. Moreover, it is one which demands passive obedience, and he implies that Christians also need to choose which Christianity they decide to follow.

Mill also raises the matter of loyalty to the state, an issue deeply divisive in many parts of Europe in Church-State relations at the beginnings of modernity. The battle between church and state in 1859 was played out in many areas, most importantly in the claims of civic authority to sovereignty, in the duties and rights of citizens, and in the control of education. Christian codes, Mill believes, are mainly silent on important matters of public authority: "What little recognition the idea of obligation to the public obtains in modern morality is derived from Greek and Roman sources, not from Christian..."

Mill and most liberals will go further. A fundamental difference between liberals (and socialists) and conservatives in the nineteenth century is in their attitude towards religion and society. Edmund Burke, in his masterful work *Reflections on the Revolution in France* (1790), a work as important to conservatism as *On Liberty* is to liberalism, attacked the revolutionaries for challenging religious authority and religious institutions. Religion, Burke claimed, was fundamental to an orderly community and is "the basis of civilized society." Hence, traditional religious institutions should be protected and it is proper for religious figures and institutions to be part of the political and constitutional structure.

Mill is suggesting that religious belief and practice is a private, not a public, matter. It is fine to have a religious belief and to practice it in assembly so long as you do no harm to others. That is a given. But he does not accept any claim that religious groups are special and should be treated differently from any other group. It is also fine, he might have said, for a group of people to found a sporting club or a chess club and to be organized. They, too, have the right of assembly. They do not have the right to demand preferential treatment.

The socialists of the mid-nineteenth century go further. Marx famously said that "religion is the opium of the people." In the *Manifesto* he and Engels claimed that law, morality, and religion are part of the superstructure used by the bourgeoisie to maintain power. Religion is treated as a political and social force, stripped of anything sacred.

The consequences of Mill's idea are great, especially in the context of his times. However, Mill here follows what would happen in France and, at least in theory, what occurred in the United States. No church or religion should be part of the polity. As we know, while the power of churches declined in the next eight score years, religions still make claims to preferential treatment and authority in many European and Western states.

A later example of what Mill called "the illegitimate interference with the rightful liberty of the individual" is Sabbatarian legislation. One person's religious duty is not another's. Certainly the idea of a day of rest is beneficial and sensible for Mill. Having to do it on Sunday and restricting all people's freedom on that day is not acceptable. He remarks in general: "I am not aware that any community has a right to force another to be civilized." He does not go so far in 1859 to discuss that what people decide is civilized might be the prejudices of a particular society. He didn't have the benefit of Gandhi's observation when asked what he thought of Western civilization. Gandhi replied that he believed it would be a good idea.

Custom, beloved by conservatives, is also challenged. A blind adherence to custom limits progress and confines a part of the population. Mill called the belief that custom was always wise and should be the basis of law and civil society a "despotism... the standing hindrance to human advancement." He understood that not all change was necessarily of value, but he was on its side. Mature individuals need to be free to make their own judgments. Becoming was beneficial to liberty and progress; Being was very limiting.

To be a liberal in the mid-nineteenth century is not necessarily to have been a democrat as we understand the term today. Mill in *On Liberty* is troubled by an aspect of modernity which was often not fully discussed in the West until after World War I by such commentators as José Ortega y Gasset. He states, "the general tendency of things throughout the world is to render mediocrity the ascendant power among mankind... [P]ublic opinion now rules the world."

Mill fears this development as he values the individual and the quest for excellence. The masses, he believes, form a "collective mediocrity." Along with Tocqueville (and Aristotle), he worries that a full democracy will result in conformity, in the social and political tyranny of the majority, which will stamp out anything other than that which resembles the mass. He argues here for diversity and variety. Europe, he believes, has prospered due to its "remarkable diversity of character and culture."

Like many who were sympathetic to workers, women and others left out of modernity, Mill wanted them as part of an active citizenry. Yet, he also feared that a full democracy could become a kind of tyranny.

Later in 1859, Mill sought something of a solution to his concerns in his pamphlet *Thoughts on Parliamentary Reform*. He suggested that a solution to the argument that the working class was not well enough educated to vote, and that they lacked property, was to consider a system of plural voting. Universal (manhood) suffrage would be introduced—property qualifications would be eliminated. However, a system of plural voting might also be implemented alongside this:

> If every ordinary unskilled labourer had one vote, a skilled labourer, whose occupation requires an exercised mind and a knowledge of some of the laws of external nature, ought to have two. A foreman, or superintendent of labour, whose occupation requires something more of general culture, and some moral as well as intellectual qualities, should perhaps have three. A farmer, manufacturer, or trader, who requires a still larger range of ideas and knowledge, and the power of guiding and attending to a great number of various operations at once, should have three or four. A member of any profession requiring a long, accurate, and systematic mental cultivation—a lawyer, a physician or surgeon, a clergyman of any denomination, a literary man, an artist, a public functionary (or, at all events, a member of every intellectual profession at the threshold of which there is a satisfactory examination test) ought to have five or six. A graduate of any university, or a person freely elected a member of any learned society, is entitled to at least as many. A certificate of having passed through a complete course of instruction at any place of education publicly recognised as one where the higher branches of knowledge are taught, should confer a plurality of votes; and there ought to be an organization of voluntary examinations throughout the country (agreeably to the precedent set by the middle-class examinations so wisely and virtuously instituted by the University of Oxford) at which any person whatever might present himself, and obtain, from impartial examiners, a certificate of his possessing the acquirements which would entitle him to any number of votes, up to the largest allowed to one individual.[54]

In the pamphlet, Mill also tried to deal with the problem of first-past-the-post voting which often eliminates representation of minorities. He suggested that each riding elect three members and that each voter have three votes, thereby giving a candidate who received all three votes from a number of voters the opportunity to be elected. Of course, Mill's proposal has its own problems, but he was far ahead of this time in realizing that proportional representation might be fairer than the current system in such places as Canada and the United Kingdom.

In *Thoughts*, Mill also argued that the vote should be given to educated householders "without distinction of sex—for why should the vote-collector make a distinction when the tax-gatherer makes none." The commentator in the *Times* derided this proposal as "doctrinaire," an "inversion of the natural order."[55]

Mill would change his position on proportional representation several times as he learned more about it. However, he never changed his views on the rights of women, arguing for their getting the franchise in parliament and in his *The Subjection of Women* (1869).

---

[54] https://en.wikisource.org/wiki/Thoughts_on_Parliamentary_Reform, paragraph 26.
[55] John Skorupski, ed., *The Cambridge Companion to Mill* (1998), 471.

Mill always was fair to his readers. The last section of *On Liberty* is titled "Applications," for he believed that acts mattered more than words. He gives several examples of how his philosophy might be used in the everyday lives of citizens.

He opens with the issue of trade, for liberalism at the time was concerned about both property and free trade. However, he argues, "trade is a social act." Contracts must be kept. Exchanges must be open between the participants. Hence, these would be other-regarding acts and are subject to public regulation.

There are, however, matters in commerce which are self-regarding and should not be regulated. On the one hand, if a person is about to commit a crime, it is reasonable to accept that public authority or private individuals can act to prevent the deed. On the other, using the sale of poisons as an example, there are limits to intervention. You can inform the person purchasing the poison of its dangers, including noting this on the label. You can register the sale. But the sale itself is legitimate. Drunkenness in private is self-regarding. A drunk who is violent to others in public may be put under certain restraints.

Then there are matters which are difficult to decide because "they lie on the exact boundaries between two principles." He cites fornication and gambling. They will happen and, though Mill does not elaborate on this, he sees them as private matters as long as there is consent among the parties concerned. However, he asks whether advising someone to do an illegal act might be other-regarding: "Should a person be free to be a pimp, or to keep a gambling house?"

Taxation to provide revenue for the state to perform its responsibilities is legitimate. However, to tax in order to make a good more difficult to obtain is not.

Mill also discusses a matter reflected upon in several works on freedom and authority: can one sell oneself, or permit oneself to be sold, into slavery? The important discussion of this issue, which Mill would have known, is that of Jean Jacques Rousseau in his *Social Contract* (1762). Rousseau summed up his position:

> To renounce liberty is to renounce being a man, to surrender the rights of humanity and even its duties. For him who renounces everything no indemnity is possible. Such a renunciation is incompatible with man's nature; to remove all liberty from his will is to remove all morality from his acts. Finally, it is an empty and contradictory *convention* that sets up, on the one side, absolute authority, and, on the other, unlimited obedience. Is it not clear that we can be under no obligation to a person from whom we have the right to exact everything? Does not this condition alone, in the absence of equivalence or exchange, in itself involve the nullity of the act? For what right can my slave have against me, when all that he has belongs to me, and, his right being mine, this right of mine against myself is a phrase devoid of meaning?

Mill is in agreement: "...by selling himself for a slave, he abdicates his liberty; he foregoes any future use of it beyond that single act." Mill appeals to the principle that each of us must have freedom in those matters which concern our own self. Self-regarding acts as part of our lives as progressive beings must be protected.

The relationship between the sexes is again raised in the applications section. Mill notes that the state has an obligation to protect the freedom of individuals to act on their own behalf. Here he moves to rights becoming far more important than any argument about utility. No person should be free to act for someone else (who is an adult) under the assumption that the affairs of the other are his own. Simply, what he calls "the almost despotic power of husbands over wives" is an "evil" which has no basis in rights and is fully about power. Women should receive the same rights and be protected by the civic authority as are men. Mill regarded this position as both logical and obvious. As well, he would argue that there would be great benefits to society if this were so. Sadly, he was correct, but not thought correct in his own time.

Mill then considered two matters which have become important in modernity in balancing authority and power, those of the importance of education and the power of the state.

He calls the belief that the state should require the education of its citizens "almost a self-evident axiom." However, the fulfillment of that requirement in 1859, he says, is left to the male parent, and if the parent does not do so, little is done by way of coercion. He states that becoming a parent means there is a responsibility to provide food for the body and also nourishment for the development of the mind. Not to do so "is a moral crime." The state must step in in these circumstances.

However, he worries about giving the state more than the responsibility of enforcing the need for universal education. Some states were, in 1859, not only requiring universal education but providing for it. The latter is anathema to Mill:

> A general State education is a mere contrivance for moulding people to be exactly like one another; and as the mould in which it casts them is that which pleases the predominant power in the government, whether this be a monarch, a priesthood, an aristocracy, or the majority of the existing generation, in proportion as it is efficient and successful, it establishes a despotism over the mind, leading by natural tendency to one over the body.

If the state does provide an education, it should only be one of many ways to obtain one. It is better for education to be both private and public. Mill proposes examinations for all children, beginning very early, as a means of determining that standards are met.

In the case of education, Mill lost. As we know, all liberal democratic states (and others) provide public education to the vast majority of their populace. The state determines standards and behavior. Most states highly regulate what is read and taught. Where there are private institutions alongside public ones, they are required to also meet state standards.

Mill's fear, however, had a basis in that he did not want the state to take over education from religious authorities and then turn the system into one of conformity to the beliefs of the state. He understood that education is also indoctrination, and feared that the state might be as bad as religious dogma in stamping out diversity.

It was Eric Hobsbawm who coined the term "the invention of tradition," and pointed to the modern state building memorials to itself and reorganizing the calendar and holidays on its own behalf. Instead of saint's days, there are now days named after presidents and dictators. Instead of religious holidays there are now celebrations devoted to the birth of modern states and important events in civic life. After all, France is an old entity, but its national day is the birth of modernity in Europe, celebrating the revolution of 1789. Similarly, bolsheviks, fascists, and democrats all manipulate public life to support their ideologies and continuity.

Mill feared state education would now help to make the state a new kind of church. And indeed, in many countries the school day no longer opens with The Lord's Prayer, but with pledges of allegiance to state authority and the singing of national anthems. The teaching of history and other subjects are highly regulated and clearly political in nature. Identity has changed, but as one reads Mill one wonders if the aim of teaching is not all that different in the modern state than it was at a time when education was under religious control.

Mill ends this part of the book with a plea for as little government interference in the freedom of individuals as possible, an argument which was prescient in these days of the surveillance state.

He lists three kinds of objections to government action. First, often "the thing that is to be done is likely to be better done by individuals than by the government." Then, even if it might be done better by government agencies, having individuals as actors rather than as passive receivers helps them to grow. But the most important reason "for restricting the interference of government is the great evil of adding unnecessarily to its power."

Mill is among the first in modernity to see that the state could become the kind of leviathan he opposed. Many political philosophers after him will deal with the issue of restraining state power. Liberals will want the state to act and fear its power at the same time. Reading Mill today, with what we know about how fascism and authoritarianism also arise out of modernity, is to nod at his worries about state power.

Similarly, before Max Weber, Mill recognized that adding to the power of government also contributed to the existence of a vast anonymous bureaucracy, which he believed was part of the apparatus of authoritarianism. Knowing what has occurred in bolshevism, fascism, and now in Chinese communism makes one read Mill on bureaucracy with much greater sympathy for his views than you might otherwise have. This is a thinker who foresaw some of the dilemmas of modernity in the political and civic spheres.

What, then, is the remedy? Mill goes to Tocqueville and his study of New England democracy for guidance:

> I believe that the practical principle in which safety resides, the ideal to be kept in view, the standard by which to test all arrangements intended for overcoming the difficulty, may be conveyed in these words: the greatest dissemination of power consistent with efficiency; but the greatest possible centralisation of information, and diffusion of it from the centre.

Centralize information and then distribute it widely. Decentralize power. Put authority in the hands of local bodies. Democracy is much more than voting for Mill. It is a way of living, whereby we have as much control as possible over the shape of our lives. Participation in the quality of daily life is highly important. The worth of a state, he concludes, "is the worth of the individuals composing it."

Liberalism absorbed much of what Mill wrote in *On Liberty* and other works. Even its contradictions and paradoxes became part of its posture. Liberals start with human rights and also have accepted the welfare (and, if necessary, the warfare) state. They emphasize free trade and argue for the regulation of monopolies and banks. They support diversity and centralized public education at the same time.

For Mill, one of the great tests of proper authority was the idea of the depoliticization of as much of life as possible. He carefully acknowledges our obligations and responsibilities to civic authority and others. However, we should be free from authority and tyranny to, insofar as it is possible, shape our lives according to our preferences and tastes. For him, there is such a thing as living democratically and supporting liberty. It includes, in addition to depoliticization, participation in civic life and being part of local decision-making. Passivity is not a part of his idea of democracy. As well, there is recognition of a shared humanity: we are not to be indifferent to the fate of our fellow citizens.

Liberalism would shift and change in the years following Mill, depending upon context and circumstances. What does happen in the West is that rights are continuously redefined and extended, especially to those left out of modernity in 1859: women, non-believers and non-Christians, people of color and those with non-heterosexual sexual orientation.

As well, liberals all over Europe, with the establishment of the EU, accept the need for a social platform as part of providing everyone an opportunity to shape one's own life. Education and healthcare are extended to all. In many places, both are added to the list of basic rights.

In a capitalist world, Liberals are "levellers." They want a much more even playing field than do conservatives, and will often do so with taxation policy. However, liberals still sometimes do not recognize that often social matters are more important in making change than political ones.

Mill continued to develop his ideas after *On Liberty*. His works on the individual and politics, on liberty and democracy, and on the dignity of all human beings, still resonate and inform. If part of modernity is the desire for all of us to be subjects rather than objects in the face of multiple pressures towards state and corporate control and treating people as things, then Mill continues to make an enormously important contribution.

Chapter VI
The New City: Manchester, Paris, and Barcelona

Europeans in the nineteenth century came up with new ideas and concepts to help them comprehend their transforming world. New terms included romanticism (c. 1800), realism (c. 1800), socialism (c. 1830), positivism (c. 1830), natural selection (1842), communism (1848), capitalism (1850), nihilism (1862), survival of the fittest (1864), ecology (1866), dystopia (1868), agnosticism (1869), impressionism (1872), anarchism (c. 1873), eugenics (1883), and the subconscious (1893).

Perhaps the term that most reflected both thought and action in relationship to the elusive concept of modernity was that of urbanism, first used in the nineteenth century by the Spanish engineer, Ildefons Cerdà (1815–1876), in 1861.

Cerdà was involved in the transformation of Barcelona in the 1850s and forward, and he struggled to understand how to develop a systematic theory and practice of urban planning. The result was the first work of a new discipline, his *General Theory of Urbanization* (1867).

Cerdà was the third son of an aristocratic family in Catalonia. He trained as a civil engineer and during his younger years he was politically active as part of the Progressive wing of Spanish liberalism. He joined the National Militia, and attained the position of lieutenant. The Militia were troops in municipalities, and, in the context of Spanish politics at the time, were seen as liberal, even radical, in their ideology. From 1850–56 Cerdà was a Deputy, an elected representative, to the Spanish parliament, a member of the Progressive Party. Later, he was a councillor in the *Consell Municipal* of Barcelona.

In 1849 Cerdà, then a government engineer, unexpectedly inherited the family estate. This enabled him to resign his position and to withdraw from the Militia, and from this time forward to pursue his great passion, the study of the modern city.

Cerdà tells of the moment he began to try to understand modernity. In 1844, he was travelling and was in the railway station in the southern French city of Nîmes, where he watched a crowd leave a train and make its way into the city:

> What surprised me, in spite of the fact that I had imagined this in my mind many times, was to see those long trains which, loaded with a large quantity of merchandise, a large number of passengers of all sexes, ages and conditions, came and went, appearing to be whole populations changing domicile. After overcoming the surprise of this spectacle, new at that time for me, my thoughts were elevated to considerations regarding the social order, specially when I observed the difficulty with which that mob of unexpected guests penetrated

through the narrow doors, scattered in the narrow streets, in search of their shelter in the mean houses of the old city."[56]

He had a eureka moment, something Baudelaire and others would have somewhat later. We are now in a time of transition, he thought: "I understood that the use of steam power to create a motive force signalled to humanity the end of an era and the start of another one."[57]

Moreover, he realized then that something never seen before was just at its beginnings and that it "will radically transform the nature and functioning of humanity." It was not just in the layout and architecture of the city that this transformation will happen. The city is the place where it will occur, but it will also take place in industry, economics, politics, and society and will be the cause of much struggle and metamorphoses.

Cerdà's insight and what he did with it parallels what will also come from Engels, Marx, and Baudelaire in a few years' time. The modern city will differ in kind from all those that called themselves "city" before the nineteenth century. It will be the landscape of the new world, ushered in by the dual revolution.

As he worked on his theory and on the transformation of Barcelona, Cerdà decided that he also needed to change the discourse about what was happening:

> ...[W]hen I decided to write something about the topic, which is as complex as it is challenging and new, the first thing that occurred to me was the need to come up with a name for this jumble of people, things, and interests of all kinds, a multitude of different elements which, although they may seems to function independently, at closer observation can be seen to exist in constant interconnection with one another, affecting one another sometimes very directly, which cements them, therefore, as a unit. We know then the sum of all these things is called a *city*... I needed to express the organism, the life, so to speak, that breathes life into the material part—it was clear I could not use the same word.[58]

He searched and found a term from the Latin language which had not entered its modern offshoots. It was *urb*, something different from a *civitas*. From *urb*, he derived the noun "urban" and the verb "urbanize" and its counterpart, as he coined it, "ruralize." And he commented in 1861 that if the modern trend is to urbanize the countryside, perhaps in thinking about the modern city, it now

---

56 Christian Hermansen, "Ildefonso Cerdá and Modernity," in *Manifestoes and Transformations in the Early Modern City*, ed. Christian Hermansen Cordua (Farnham: England, Ashgate), 223.
57 Ibid.
58 Ildefons Cerdà, *General Theory of Urbanization* (Barcelona: Institute for Advanced Architecture of Catalonia, 2018), 66–67.

needs to be ruralised.⁵⁹ Cerdà (and a number of commentators on Cerdà) seems to have been unaware that the term *urbs* was used in Spain in the discussion of cities as far back as Isidore of Seville (560–636) who distinguished between a *civitas*, a social body, and an *urbs*, an architectural designation. As well, the term was used in the sixteenth and seventeenth centuries by several commentators to indicate what Kagan calls a "topographical" entity.⁶⁰

Around the year 1859 there were three European cities commonly considered at the forefront of how the transformation to modernity was occurring. They were Manchester, the industrial city *par excellence*; Paris, being modernized by the government and viewed at the time as the capital of Europe; and Barcelona, an old city which tore down its walls and came to be in the forefront of the new city planning and, later in the century, exemplified a new *Modernisme* style.

Manchester

No city reflected the possibilities and horrors of the new industrial age more than Manchester, located in the county of Lancashire in the northwest of England, roughly 200 miles from London and 30 miles from its port of Liverpool.

The early industrial revolution was based upon textiles. Manchester had a cotton trade in the eighteenth century, and the city then became the beneficiary of changes in production brought about by new inventions—the spinning jenny (1764) and the spinning mule (1775–79), machines used to spin fibres. However, the most transformative development in the late eighteenth century in changing the way people lived was the steam engine, patented by James Watt in 1769, the first rotary steam engine being put to use in 1784. Now there were mechanical means of manufacturing and a certain and cheap source of power. By the end of the Napoleonic wars, there were immense spinning mills in the north of England, centered on the town of Manchester.

The population of Manchester increased from 24,000 (estimated) in 1773 to 70,000 as determined by the first census of 1801. It doubled by 1831 and tripled by 1841. It was a bit over 250,000 by 1851. It spread out quickly and took in the community of Salford on the other side of the river Irwell.⁶¹

---

59 *Cerdà: The Five Bases of the General Theory of Urbanization*, ed. Arturo Soria y Puig (Barcelona: Elect, 1999), 87.

60 Richard L. Kagan, "*Urbs* and *Civitas* in Sixteenth- and Seventeenth-Century Spain," in *Envisioning the City*, ed. David Buissert (Chicago: University of Chicago Press, 1998), 75–108. Cerdà's use of the term *urb* includes both society and topography.

61 Steven Marcus, *Engels, Manchester and the Working Class* (New York: Routledge, 2015), 4.

Manchester's initial growth occurred randomly, with no plan and no supervision. It grew so quickly in a time of laissez-faire that little was overseen. Manchester was still legally a manor in the early nineteenth century, essentially a carryover from medieval times. It had few civic institutions, little policing, an archaic court system, and did not elect any representatives to Parliament.

The Reform Bill of 1832 gave some of Manchester's middle class the vote and the town now was designated as a borough which elected two members of Parliament. It was incorporated in 1838 and finally had a council, which in the 1840s attempted reform. In 1845, the municipality purchased the manorial rights from the Mosley family, giving it more authority. It became an incorporated city in 1853.

Hence, in the first half of the nineteenth century a place which had fast become one of the wealthiest areas in the world had little or no government. The town, the world center of the cotton industry, which imported cotton from the southern United States, Asia, and the West Indies, and which shipped finished products worldwide, simply grew according to the needs of the manufacturing and middle classes. Liverpool became the cotton port, necessary to Manchester's distribution. British cotton exports were worth £1,000,000 in 1785; by 1851 they were worth £31,000,000.[62]

The introduction of the railway acknowledged the importance of this commerce. The first passenger railway connection was Manchester-Liverpool, completed in 1830, using locomotives driven by steam power. Manchester became a railway center for all of the many manufacturing towns in the north. It was then linked to London via Birmingham. Soon there were four railway stations in the town. Asa Briggs stated that by 1840 "Manchester was far more than a 'metropolis of manufactures': it was above all a center of trade of a whole region, linked with the whole world."[63]

The unprecedented growth created conflict and resentment in a good deal of the population. Working conditions were terrible. Living conditions were horrific. Manchester soon became a place where the new industrial society was evaluated and critiqued.

Labor strife was endemic, though labor had few legal alternatives. There was a strike of cotton spinners in 1810 and in 1818 there was an effort to establish a "general union of trades." In the decade 1810–20 there were demonstrations, rioting, and the breaking of machines by those known as Luddites. Owenites opened an Owenite Hall of Science in 1839 as a place to further the socialist

---

[62] Ian Taylor et al., *A Tale of Two Cities* (New York: Routledge, 1996), 50.
[63] Asa Briggs, *Victorian Cities* (London: Oldham Books Ltd., 1963), 102.

ideas of Robert Owen, who had managed cotton factories in Manchester in the late eighteenth century, and whose communal ideas were popular. Various skilled trades established clubs, harking back to their past as guilds. There were regular street riots. Mancunians were central in Chartist attempts at reform.[64]

The event that made the social and political issues around the industrial revolution a national concern occurred in 1819. On August 16, 60,000 people assembled at St. Peter's field in Manchester. The purpose of the gathering was announced as "to take into consideration the most speedy and effectual mode of obtaining Radical Reform in the Commons House of Parliament; and to consider the propriety of the unrepresented inhabitants of Manchester electing a person to represent them in Parliament."[65]

The main speaker was to be a reform leader, Henry Hunt. Constables were present to maintain order, with infantry and mounted yeomanry ready outside the crowd. The authorities were frightened by the mood of the crowd and ordered Hunt arrested. Confusion and chaos occurred as constables and mounted troops entered the crowd. The result was violence and panic. When it was over some were killed (it is not clear how many; estimates are as low as six and as high as 15) and several hundred were injured.

The incident became a famous moment, reported to the world. Recalling the battle of Waterloo only four years previously, the incident became Peterloo and the Peterloo Massacre soon became mythologized. People wore scarves with the name and images of the event, and bought pottery and mugs in commemoration.

The *Manchester Guardian* (now *The Guardian*) was founded in the wake of Peterloo to represent the liberal viewpoint in the city and to give a voice to those left out of the social and political discourse at the time. It published its first daily paper in 1821. In 1832, Peterloo was still a reference point, as the newspaper quoted a popular parody on the regular prayer:

> From all those damnable bishops, lords and peers,
> from all those bloody murdering Peterloo butchers –
> from all those idle drones that live out of the earning
> of the people – Good Lord deliver us.[66]

---

[64] W. H. Chaloner, "The Birth of Modern Manchester," *Manchester and Its Region* (Manchester University Press, 1962), 136.
[65] F.A. Bruton, *A Short History of Manchester and Salford* (East Ardsley: S. R. Publishers Limited, 1970), 172.
[66] Gary S. Messinger, *Manchester in the Victorian Age* (Manchester University Press, 1985), 30.

Fig. 12: *Peterloo Massacre*, Print published by Richard Carlile, October 1, 1819

Manchester became the emblematic industrial city in the first generation of the mechanization of production, the growth of the factory, and the conflict between the classes. For the manufacturers it was a place to make money. For the laborers it was hellish and unfair.

Other matters of importance were also part of Manchester's history before the middle of the century. It was there that the middle-class defended the doctrine of free trade and laissez-faire (though those who defended laissez-faire agitated for roads, railways, police protection, and municipal services from the government). They were especially opposed to the Corn Laws, the first of which was passed in 1815, which controlled the cost of grain and favored the landowners over the industrial owners and workers. The Manchester Anti-Corn Law Association was founded in 1838, becoming a national body, the Anti-Corn Law League, the next year. Their active efforts were successful in 1846, with the repeal of the legislation.

Manchester was also a center for those who opposed the New Poor Law of 1834, and for attempts to legalize unions and reform factory labor. Many cooperatives and trade union groups existed. When the Chartist movement began in the late 1830s Manchester became the center of the movement in Lancashire. Though the aims of the Chartists' "six points" were political (universal manhood suffrage, pay for members of Parliament, annual elections, secret ballot, fair election districts, and an end to the property qualification for members of parliament), the movement was driven by economic and social forces, and it hoped for a fairer society and an amelioration of the plight of the British laborer.

Manchester soon became a place to study and visit for those trying to comprehend the rapid transformation to modernity in the early nineteenth century. In 1835, the year the first volume of his monumental *Democracy in America* was published, Alexis de Tocqueville visited the city. He first commented on what favored the town: its location near the great port of Liverpool and the nearby coal mines to keep the machines, made locally or only several miles away, working; the canals and the railway; the vast numbers of laborers available in the area; the use of science and the growth of capital. Then:

> Thirty or forty factories rise on the tops of the hills I have just described. Their six stories tower up; their huge enclosures give notice from afar of the centralization of industry. The wretched dwellings of the poor are scattered haphazards around them. Round them stretches land uncultivated but without the charm of rustic nature, and still without the amenities of a town… The land is given over to the industry's use. Heaps of dung, rubble from buildings, putrid, stagnant pools are found here and there among the houses and over the bumpy, pitted surfaces of the public places. No trace of surveyor's rod or spirit-level. Amid this noisome labyrinth, this great, sombre stretch of brickwork, from time to time one is astonished at the sight of fine stone buildings with Corinthian columns. It might be a medieval town with the marvels of the nineteenth century in the middle of it. But who could describe the interiors of these quarters set apart, home of vice and poverty, which surround the huge palaces of industry and clasp them in their hideous folds.... [In some dwellings] twelve to fifteen human beings are crowded pell-mell into each of these damp, repulsive holes… It is [a] new Hades. A sort of black smoke covers the city. The sun seen through it is a disc without rays. Under this half daylight 300,000 human beings are ceaselessly at work. A thousand noises disturb this damp, dark labyrinth… The footsteps of a busy crowd, the crunching wheels of machinery, the shriek of steam from boilers, the regular beat of the looms, the heavy rumble of carts, those are the noises from which you can never escape in the sombre half-light of these streets. From this foul drain the greatest stream of human industry flows out to fertilize the whole world. From this filthy sewer pure gold flows. Here humanity attains its most complete development and its most brutish; here civilisation works its miracles, and civilised man is turned back almost into a savage.[67]

The next decade would see the most famous of all commentators on industrial Manchester, Friedrich Engels (1820–1895), come to live in Manchester from the end of November 1842 to mid-1844. He was sent to work in the family firm, the Ermen and Engels' Victoria Mill, with the hope that he would learn

---

[67] Alexis de Tocqueville, *Journeys to England and Ireland* (New Haven, CT: Yale University Press, 1958), 104–7.

about commerce and hopefully moderate his radical views. He did learn about commerce. However, his experience of the city only cemented his radicalism.

Engels' study, *The Condition of the Working Class in England*, is an extraordinary accomplishment for a 24-year-old who was mainly self-educated. He wrote well and combined careful observation, statistics, interesting examples, a sense of moral concern, and an analysis of class in his work. He was writing at about the same time as Karl Marx, who he had met once on his way to Manchester, though the two did not become friends until after his stay in the city. Marx was writing what we have come to call his *Economic and Philosophic Manuscripts of 1844*, not published until 1927 in Moscow and 1932 in Germany. There are many striking similarities in the analyses of class conflict and the proletariat in both works. It is easy to understand the collaboration which, in 1848, would result in the *Manifesto of the Communist Party*.

Manchester is first discussed by Engels as a place where the "middle class," as he calls the owners of industry, sometimes referring to them as the "money aristocracy" (Marx, and soon Engels, will call them the bourgeoisie), and the industrial proletariat are completely separated in their living quarters, so much so that the wealthy can go to their businesses "without ever seeing that they are in the midst of... grimy misery... which form the complement to their wealth."[68]

Engels then describes an area of the city which is working-class, letting the description of the houses, the housing in cellars, the filth and overcrowding, the lack of health provisions, etc. do the work for him. He concludes: "In a word, we must confess that in the working men's dwellings of Manchester, no cleanliness, no convenience, and consequently no family life is possible." Indeed, there was a cholera epidemic that took hold in Manchester in 1832. Engels notes that on one street which followed a filthy ditch, not a single house escaped the disease.

He also discusses such things as diet and clothing, noting that as the wages of workers decline, people live on such things as bread, cheese, porridge and, most of all, potatoes. Further, there are times when there is no work for some: "[W]hen he has [no work], he is wholly at the mercy of accident, and eats what is given him, what he can beg or steal. And, if he gets nothing, he simply starves...."

The study uses whatever reliable statistics Engels can gather. Among other information, he notes that in Liverpool in 1840 the average life expectancy of the upper classes is 35; for the day-laborers, operatives in factories and those in service it was 15. In Manchester 57 percent of the children of the proletariat

---

68 Friedrich Engels, *The Condition of the Working Class in England* (London: Penguin Books, 2005), 86.

die before they are five years old. Death rates in cities in the north of England are far higher than those living in rural areas. He even investigates the weight and height of the people he is studying.

There is moral outrage in Engels. There are many passages like the following: "...[T]he English middle class, especially the manufacturing class, which is enriched directly by the poverty of the workers, persists in ignoring this poverty." The book is a fine work of sociology (before there was anything like the discipline of sociology), but it is also a reflection in moral philosophy. Like many novelists, he knows something is very wrong in this town and in the system. Unlike some of them, he would not be upset by a total change in the way politics and society are organized.

Near the end of the study Engels relates an anecdote, an experience he had, which stayed with him. It is a passage much quoted by commentators on the ravages of the industrial revolution and the horrors seen daily in early industrial cities:

> I have never seen a class so deeply demoralized, so incurably debased by selfishness, so corroded within, so incapable of progress, as the English bourgeoisie... I once went into Manchester with such a bourgeois, and spoke to him of the bad, unwholesome method of building, the frightful condition of the working people's quarters, and said that I had never seen so ill-built a city. The man listened quietly to the end, and said at the corner where we parted: 'And yet there is a great deal of money to be made here; good morning, sir.'

At the same time, in 1844, Marx, 26 years old, is writing about the dilemmas presented by both property and money as it is used in industrial society. He quotes Goethe, who died in 1832, from *Faust*, a work which, among many matters, also deals with the moral consequences of construction and destruction in the modern era. Mephistopheles speaks as he is making his bargain with Faust:

> What, man! Confound it, hands and feet
> And head and backside, all are yours!
> And what we take while life is sweet,
> Is that to be declared not ours?
> Six stallions, say, I can afford,
> Is not their strength my property?
> I tear along, a sporting lord,
> As if their legs belonged to me.

Like many important cities, Manchester became the subject of novelists. The first major industrial city, it spawned a new genre, the industrial novel, sometimes referred to as Condition-of-England novels.

Disraeli's *Sybil, or the Two Nations* (1845) is among the first novels to address the problems of both urban and rural poverty in England after the passage of the Reform Bill in 1832. The fictional town of Mowbray, a composite of industrial towns in the north, most nearly resembles the Manchester of Engels. The two nations, for Disraeli, are the rich and the poor: "Two nations between whom there is no intercourse and no sympathy; who are as ignorant of each other's habits, thoughts, and feelings, as if they were dwellers in different zones, or inhabitants of different planets." The goal is reform and reconciliation.

Elizabeth Gaskell's two popular and important works, *Mary Barton* (1848) and *North and South* (1854), are set in Manchester. Gaskell was the wife of the Reverend William Gaskell, a Unitarian minister in the city, and she knew the issues she discussed from her daily life. There are long and insightful descriptions of domestic life and class divisions in both books, including a lament about the hypocrisy of the rich. John Barton is determined that his daughter Mary "shall never work in a factory." And then, in a conversation with a friend, he says,

> If I am sick, do they [the middle class, called the 'gentlefolk'] come and nurse me? If my child lays dying does the rich man bring the wine or broth that might save his life? If I am out of work for weeks in the bad times, and winter comes, with black frost, and keen east wind, and there is no coal for the grate, and no clothes for the bed, and the thin ones are seen through the ragged clothes, does the rich man share his plenty with me, as he ought to do, if his religion wasn't a humbug?... No, I tell you, it's the poor, and the poor only, as does such things for the poor... We're their slaves as long as we can work. We pile up their fortunes with the sweat of our brows...[69]

Gaskell was especially concerned with the status of women in industrial society. She realized that working class women had very few alternatives, nearly all of them horrific—work in a factory or turn to prostitution. She pleaded for giving agency to women in a masculine culture.

Charles Dickens' *Hard Times* (1854) takes place in Coketown, a fictional Manchester:

> It was a town of red brick, or of brick that would have been red if the smoke and ashes had allowed it; but as matters stood, it was a town of unnatural red and black like the painted face of a savage. It was a town of machinery and tall chimneys, out of which interminable serpents of smoke trailed themselves for ever and ever, and never got uncoiled. It had a black canal in it, and a river that ran purple with ill-smelling dye, and vast piles of building full of windows where there was a rattling and a trembling all day long, and where the pis-

---

[69] Elizabeth Gaskell, *Mary Barton* (Peterborough, Canada: Broadview Press, 2000), 40.

ton of the steam-engine worked monotonously up and down, like the head of an elephant in a state of melancholy madness.⁷⁰

In an earlier novel, *Bleak House* (1853), Dickens deals with matters about the law, the law courts, and the city of London. In the second paragraph of the work, the word "fog" is used 13 times to deal with both the obfuscation of the law and the atmosphere in the city.

In *Hard Times*, Dickens has no sympathy for the utilitarian rationalist, Gradgrind, the vile manufacturer, Bounderby, or the ambitious union leader, Slackbridge, as their names attest. Still, he does care about the poor and his hero is Stephen Blackpool, a very decent man, a laborer caught in a time when he had no choices, or as Dickens put it in a chapter heading, "no way out."

All of the English social novelists pleaded for reform, not revolution, hoping that the wealthy would recognize their social responsibilities and behave decently. Though they are in agreement about the problems of the industrial city, there is a vast gap between Engels and the others in terms of the solutions.

Paris

Paris in 1859 was in the midst of a vast transformation from a city with medieval streets to one which would reflect the new world of Becoming in the nineteenth century. Far from letting it happen randomly, it was the government of Napoleon III which mandated the changes and it was Baron Georges-Eugène Haussmann who became the first of the modern developers, with the extraordinary power to both destroy and construct almost at will. Baudelaire was the city's poet, Daumier was its caricaturist, Manet was its artist, and Haussmann was its builder.

Napoleon III had been elected president of France in late 1848 for a four-year term. Like his uncle Napoleon Bonaparte, who became Napoleon I, he pulled a coup on his own state in 1851 and repressed opponents. Again emulating his uncle, in December 1852 he held a referendum on his new authority and on his becoming emperor. He won overwhelmingly, and the Second Empire of France was established, lasting until France's defeat in the Franco-Prussian War in 1870.

He called himself Napoleon III, recognizing that Napoleon I had a son who died in 1832, who had styled himself as Napoleon II. His full title was Napoleon

---

70 Charles Dickens, *Hard Times* (New York: W. W. Norton and Company, 1990), 20.

III, Emperor of the French (not Emperor of France), and like his uncle, he claimed his legitimacy from the people. By now, France provided right-wing dictators with a model that would be used by many in the twentieth and twenty-first centuries.

Napoleon was a follower of St-Simon in that he wanted the "new men," the scientists and the bourgeoisie, to be active in power. He also expected that the state would become a builder—totally unlike what happened in early Manchester. One of his legacies would be the rebuilding of Paris in order to bring it into modernity. "We have," he stated in October 1852, "immense uncultivated lands to clear, harbours to excavate, rivers to make navigable, canals to finish, our railway networks to complete."[71] To that end, in 1853, he appointed George-Eugène Haussmann as the prefect of the Seine, in effect the civil servant responsible for the entire project.

Haussmann was a Parisian and had much experience in the development of urban areas. In his early career he was an under-prefect in many localities all over the country. He rose to become a prefect and in 1851 Louis-Napoleon appointed him prefect of the Gironde, centered in Bordeaux, where he was highly successful. Dissatisfied with the current prefect of the Seine, Jean-Louis Berger, Napoleon elevated Haussmann to Paris.

Napoleon and Haussmann were in agreement on the general goals of the transformation. Most important, they started with the idea of turning what were medieval streets into broad modern boulevards, linked by monuments. There was to be an axial symmetry, for example the extension of the Rue de Rivoli to the Hôtel de Ville—which resulted in the destruction of some of the oldest areas of the city, themselves small neighborhoods—and the Louvre-Tuileries axis, linking the Place de la Concorde to the Arc de Triomphe. Haussmann's Paris was, as one commentator has noted, "a city of movement between zones defined by function, class, and activity, and delimited by boulevards."[72]

As Haussmann stated in his memoirs:

> It was a great satisfaction for me to start out in Paris by razing [the shoddy neighborhood around the Louvre]. Since my youth, the uncared-for state of the Place du Carrousel in front of the courtyard of the Tuileries seemed to me to be a shame for France, an avowal of its government's powerlessness, of which I was resentful.[73]

---

[71] David Harvey, *Paris, Capital of Modernity* (New York: Routledge, 2006), 107.
[72] David P. Jordan, *Transforming Paris: The Life and Labors of Baron Haussmann* (Chicago: University of Chicago Press, 1995), 175.
[73] Stephane Kirkland, *Paris Reborn* (New York: St. Martin's Press, 2013), 90.

Neither Napoleon nor Haussmann were patient with democratic norms or processes. There were municipal institutions. However, both men overrode them if necessary.

Haussmann, supported and encouraged by Napoleon, was thus among the first of the modern developers identified by Marshall Berman in his insightful work *All That is Solid Melts into Air*. As Berman noted, such persons were both destroyers and builders. Paris was a construction site for most of Napoleon's reign. Napoleon himself was trying to be both his uncle and Augustus, leaving behind a legacy which included modern urban development, new monuments, and a grandeur that reflected his reign. Happily, he avoided the brutalism of many of the dictators and fascists who would follow him.

There was another change that Haussmann introduced, relating to the financing of this massive project. He was among the first to put into practice the idea that debt financing of new modern infrastructure would be positive in that it would yield new revenues, as well as result in higher valuations of land and real estate, which would soon pay back the debt. The economy would rise and all would be well. This idea ran counter to the laissez-faire attitudes and the fiscal conservatism of many of the bourgeois class, which saw debt as a sign of undisciplined management. However, for Haussmann "the city was a revenue base to be managed as an asset." Development meant linking areas to make for greater commerce and efficiency.[74]

There was yet another purpose that many believed to be a part of this revision of the city. France, and Paris especially, already had a tradition of street demonstrations, riots, and revolts. Indeed, in the old Paris of tiny winding streets it was common for demonstrators to erect barricades to control an area and to challenge authority. Hence, many believed (and still do) that the plan was also designed to permit the government to have greater control of urban unrest. Haussmann remarked:

> We ripped open the belly of Paris, the neighborhood of revolts and barricades, and cut a large opening through the almost impenetrable maze of alleys, piece by piece, and put in cross-streets whose continuations terminated the work... In putting in the boulevard de Strasbourg, and extending it to the Seine and beyond, I am certain the Emperor did not have its strategic usefulness in mind;... But even if this was not his specific intention, though the opposition kept accusing him of it, there is no denying that it was the most opportune result of all the large thoroughfares... As for myself... I can guarantee that their greater or lesser strategic importance was farthest from my mind.[75]

---

74 Kirkland, op. cit., 80.
75 Johannes Willms, *Paris, Capital of Europe* (New York: Holmes and Meier, 1997), 266.

Paris witnessed revolution in the streets and changes in the government of France in 1789, 1792, 1830, and 1848. As well, there had been many smaller uprisings and demonstrations. However, even after the transformation, especially in 1968 and 2019, the city still could express dissent with force on the new boulevards.

Many important new projects became part of the metamorphosis. Public parks were introduced and developed, most notably the Bois de Boulogne in the west and the Bois de Vincennes in the east.

The area of the Bois de Boulogne had been a forest, sometimes used by royalty in the *ancien régime* as a hunting ground. At the end of the Napoleonic wars in 1815 it came to be used briefly as a camping ground for Russian and British soldiers. Trees were cut down to build shelter for the soldiers and to obtain firewood for warmth. After they left, the area was dismal, full of neglected meadows, ponds that were filthy, and many tree stumps. The state owned the land and in 1852 Napoleon III had the whole of it, 845 hectares, two and half times the size of Central Park in New York, ceded to the city of Paris.

Napoleon III got the idea of a great public park from his knowledge and admiration of London's Hyde Park. He initially wanted it to have a watercourse similar to that of London's Serpentine. However, the initial attempt to create this was blundered by Berger and the project seemed to be a failure. Haussmann, replacing Berger in 1853, managed to rescue the matter by creating an upper and a lower lake, with water from the Seine flowing across an artificial canal to supply both lakes and other streams, lakes and a waterfall.

This success gave Haussmann the opportunity to go further with the project. He brought in two specialists with whom he had worked in Bordeaux, an engineer, Jean-Charles Alphand, and a landscape architect, Jean-Pierre Barillet-Deschamps. Alphand took the flat meadows and turned them into a landscape of great variety, including undulating hills, islands, and paths that were curved and suited for interesting walks. 420,000 trees were planted, lawns were created, flowers placed in many areas.

The park was not just conceived as a wonderful place to be in nature. It was also to be a place of sporting fields, cafes, bandstands, riding stables, boating on its lakes, and restaurants. On the south end of the park, the Longchamp racetrack opened in 1857. On the north end there was a small zoo.

The Bois de Boulogne was on the west end of Paris, the well-off part of the city. For the more working-class east end, Napoleon III conceived of taking the military training ground that was in the Bois de Vincennes and also making it into a great park, slightly larger than the Bois de Boulogne. The same team worked on it, and in short order it had its own set of lakes, islands, greenery, walks, cafes, and amusements, including a race-track. The two parks came to be called "the lungs of Paris."

Smaller parks were also included in the redoing of the city, places where urban strollers could sit, rest, have something to eat, take children and simply observe the passing scene. These were complemented by squares placed in many neighborhoods.

The two large park projects reflected what was happening socially in Paris as a result of the transformation. The old Paris was heterogeneous in terms of where the classes lived and shopped and played. In apartments, the wealthy lived on the lower floors. The further up one went, the lower the class in the social structure, with the poor living in the attic. Honoré de Balzac (1800–1850) records all this in his monumental set of novels collectively known as *La Comédie Humaine*.[76]

The new Paris would put the wealthy with their own, and the poor with the poor. It is agreed by scholars that while the rich and the middle-class benefitted greatly from the changes from 1853–1870, "the failure to improve sufficiently the living conditions of the Parisian working class… was a great missed opportunity of Second Empire policy." Workers' housing was not much improved and rents rose during the period. The *rentier* group of the bourgeoisie did well; their tenants were not protected. The working class came to live on the outskirts of the city or in overcrowded neighborhoods in the fourth and fifth arrondissements.[77]

The gulf between the classes grew even greater and so did the hostility between them. Paris was made very beautiful in much of its center and inner ring, but for many on the outskirts this was not the case. Max Nordau (1849–1923) described Belleville, a working-class district, on the edge of Paris. It echoes some of what was said of Manchester:

> The houses, located along random, dirty, goat path-like alleys, look neglected and flimsy; the plaster is peeling off the dirty fronts; the few stucco ornaments on the rafters are crumbling and fall off piece by piece; the gate hinges are loose; the holes in the windows are covered with waxed paper. The whole impression is one of walking between two rows of ragged beggars, held together with string and adhesive tape, the mere sight of whom is an unexpressed plea for alms. There are not many stores.… But almost every building contains a brandy or wine store and a soup kitchen that looks most repulsive.[78]

---

76 See the opening of Balzac's *Père Goriot*, for example.
77 Willms, op. cit., 287.
78 Willms, op. cit., 288.

Fig. 13: Honoré Daumier, *Tenants and Landlords*, published in *Le Charivari*, February 21, 1854. M. Vautour ._ "Good!.. another house is being razed...I can raise the rents of each of my lodgers by two hundred francs!" The landlord, M. Vautour, contemplates the new Paris. M. Vautour became the stock rapacious landlord of the time, even being portrayed and satirized onstage.

Ever since the great revolution of 1789, there has been much speculation on who was the victor in modern France. For the years 1789–1870 there is no doubt that it is the bourgeoisie, as seen through the eyes of such perceptive commentators as Saint-Simon, Tocqueville, Balzac, Flaubert, Baudelaire, Marx, and Engels. Paris came to be regarded by many as the most beautiful city in the modern world. As Baudelaire remarked in 1846, it was also the city of the bourgeoisie: "You [the bourgeoisie] are the majority in numbers and intelligence; therefore you are the force—which is justice... The government of the city is in your hands."

There were parks, leisure, and amusements in the new city. However, there was also attention paid to public health. Paris itself, like many cities in the mid-nineteenth century, smelled. Human waste was collected at night by private companies, shovelled into barrels and taken by carts to storage tanks located in the north of the city, then taken to a forest to be dumped. The malodorous waste left its traces in the air.

A new sewer system was a necessary part of the plan. Haussmann did some investigation and rejected the model used by the English which sent solid waste, household waste, and rainwater into the same system. Waste would be collected by one sewer network; other products and water by another, made up of smaller sewers located in neighborhoods. The network measured 66 miles in 1854. Finished, it was more than five times that size. Haussmann recognized in his memoirs that this was no small accomplishment. Building the new sewer system was the "greatest, most meritorious, if not the most spectacular and most appreciated, service that an administration conscious of its duties can provide the population."[79] Cleanliness, remarked many in Europe, including Ruskin and, later, Freud, came to be understood as one of the foundations of a civilized life.

Paris' drinking water at the start of the Second Empire was mainly from the Seine, brackish and full of solids, quite disgusting to most. As well it was a main cause of disease, especially to foreigners who travelled to the city. There had recently been two cholera epidemics. The first, in 1832, killed 20,000; the second, in 1849, killed 19,000.[80]

The supply of water to the city was changed under Haussmann. A plan was developed for reservoirs and for an aqueduct to bring clean water to homes from Champagne, 60 miles to the north. A private company was hired to organize the distribution.

France had been centralizing for at least a century and a half, back to the time of Louis XIV. It was Tocqueville who in 1856 published his insightful study, *The Old Regime and the French Revolution,* in which he argued for continuity between the public authorities prior to 1789 and after in terms of centralizing power, even sometimes using autocratic methods to do so. This trend is also seen in the administration of the new Paris in developments occurring administratively in the Île de la Cité and the distribution of food in creating Les Halles.

The Île de la Cité, the larger of the two islands in the middle of Paris and in the Seine (the smaller is the Île St. Louis, which remained a residential area), housed Notre-Dame, an old hospital, the Palais de Justice, including Sainte-Chapelle, and a lot of poor housing. The housing was razed and the island

---

[79] Kirkland, op. cit., 117.
[80] Patrice Higonnet, *Paris, Capital of the World* (Cambridge: Harvard University Press, 2002), 183.

was turned into a center for the administration of the city. Its residential population declined by two-thirds, from 15,000 to 5,000. The hospital was torn down and another was built near Notre-Dame. In the middle of the island there was now the Préfecture de Police and the commercial courts. Many came to visit Notre-Dame, part of the soul of France and Europe. Fewer visited the beautiful Sainte-Chapelle, once the chapel of the kings of France. Otherwise, the island became both the administrative center of the city and a place to contemplate the Seine and to move between the left and right banks of the river.[81]

Modernizing and centralizing the distribution of food was as important to Napoleon III and Haussmann as were the need for sewers and good drinking water. Napoleon wanted the design for Les Halles to reflect the same kind of modernity that had guided the now famous Crystal Palace in London in 1851. Nothing classical, nothing heavy. Rather, a metal structure. The first plans were rejected and Haussmann redid them. He instructed the architect, Victor Baltard, "Iron, iron—nothing but iron."

The result was one of the first modern structures using metal and glass and the reconstruction of one of the oldest neighborhoods of the city. Over 300 houses were demolished. Even the roads around Les Halles were reorganized to link it to the main boulevards.[82]

The area came to be known as *Le Ventre de Paris*, the Belly of Paris, and that is the title of an 1873 novel by Émile Zola, the third in his 20-volume series *Les Rougon-Macquart*. It has been translated into English as *The Fat and the Thin*. He described the new market:

> During the long rambles when Claude, Cadine, and Marjolin prowled about the neighbourhood of the markets, they saw the iron ribs of the giant building at the end of every street. Wherever they turned they caught sudden glimpses of it; the horizon was always bounded by it; merely the aspect under which it was seen varied. Claude was perpetually turning round, and particularly in the Rue Montmartre, after passing the church. From that point the markets, seen obliquely in the distance, filled him with enthusiasm. A huge arcade, a giant, gaping gateway, was open before him; then came the crowding pavilions with their lower and upper roofs, their countless Venetian shutters and endless blinds, a vision, as it were, of superposed houses and palaces; a Babylon of metal of Hindoo delicacy of workmanship, intersected by hanging terraces, aerial galleries, and flying bridges poised over space... They came back to it during the hot afternoons when the Venetian shutters were closed and the blinds lowered. In the covered ways all seemed to be asleep, the ashy greyness was streaked by yellow bars of sunlight falling through the high windows. Only a subdued murmur broke the silence; the steps of a few hurrying passers-by resounded on the footways; whilst the badge-wearing porters sat in rows on the stone ledges at the

---

81 Colin Jones, *Paris: Biography of a City* (London: Allen Lane, 2004), 354.
82 Michel Carmona, *Haussmann*, (Publishers Weekly, 2002), 183.

corners of the pavilions, taking off their boots and nursing their aching feet. The quietude was that of a colossus at rest, interrupted at times by some cock-crow rising from the cellars below.

He was fond, too, of the footways of the Rue Rambuteau and the Rue du Pont Neuf, near the fruit market, where the retail dealers congregated. The sight of the vegetables displayed in the open air, on trestle-tables covered with damp black rags, was full of charm for him. At four in the afternoon the whole of this nook of greenery was aglow with sunshine... however, on the opposite footway... he found a splendid subject for a picture in the stall-keepers squatting under their huge umbrellas of faded red, blue, and violet, which, mounted upon poles, filled the whole market-side with bumps, and showed conspicuously against the fiery glow of the sinking sun, whose rays faded amidst the carrots and the turnips.[83]

And Paris expanded. In 1860 what were thought of as communes related to the city were incorporated into Paris. The city took in 11 full communes, including Belleville and Montmartre, and parts of 13 others. Its area increased by two and a half times, to 33 square miles, and it added 400,000 people, bringing its total population to 1.5 million. A new arrondissement arrangement was established, one which rationalized the administration of the area.

It was not uncommon for large and powerful cities to expand their control in the mid-nineteenth century. The areas on their traditional boundaries were becoming part of their economic and social reality. Hence Manchester and Salford virtually became one; London absorbed small neighborhoods and villages on its borders; and Barcelona, as we shall see, tore down its walls and increased its size considerably. The new powerful city—be it our three under consideration, New York, London, Vienna, Berlin, St. Petersburg, and others—was at the heart of what modernity was about and where it was situated.

The number of buildings constructed in the renovation of Paris contributed to a uniformity in height and style which continues to give the city some of its personality. Many of the six-story buildings were rentals, and they were designed to house the growing bourgeois class.[84] It was required that the buildings have similar facades and more or less the same height.[85] The lines of the balconies and architectural decorations often continue for a block, sometimes more. It gives Paris streets a recognizable style, unlike many cities where developers were permitted a good deal more freedom.

The period of the Second Empire thus saw 85 miles of new streets built, most far wider than what had existed before. It was the era of gas streetlamps, provid-

---

83 http://www.gutenberg.org/files/5744/5744-h/5744-h.htm.
84 Willms, op. cit., 272.
85 Kirkland, op. cit., 141–42.

ing an opportunity for night life never experienced before. The number of streetlamps more than doubled in the Haussmannian years. As in the public parks, trees became an important part of the city. By 1870 there were nearly 100,000 along the various streets and roads.[86] Many remarked during the period that Paris was both a city and a construction zone. As a result of all the building and the urban segregation of the classes, roughly 350,000 people were displaced during the period.[87] Lives changed. Such massive change could only be accomplished in a period when there were few restraints on the power of government to confiscate land in the name of progress. It, too, became a model for later authoritarian governments, even some democratic ones.

The whole experience of space and time was transformed in the new modern Paris. As Harvey remarked, the city became a place that operated "at a different speed and scale." As well, the city was now a place of spectacle, with its boulevards, entertainments, department stores, and cafes.[88]

The most singular building and neighboring area which came to be seen as a reflection of the new Paris was the new Paris Opera and the area around it. In 1860 Napoleon announced that a competition would be held for the design of a new theater for opera, a reflection of the culture of France and European civilization, just as Les Halles, railway terminals, and gas lit streets were symbols of industrial progress.

There were 170 (some say 171) submissions to the jury. The winner was the design of an architect not well known at the time, Charles Garnier. Modern techniques were to be used—its structure was to be of iron—but this was to be a building harking back to tradition in its look. It has elements of the Baroque and Classicism, as well as the Renaissance. It is symmetrical. The mosaics celebrate earlier musicians and poets. A statue of Apollo is on the high point of its roof.

When Garnier was notified that he had won the competition, he was told to go to the Tuileries to meet Napoleon III and the Empress Eugénie and to present the project to them. The Empress was hostile, because the jury had not chosen her preferred architect, the well-known Viollet-le-Duc. It is said that she told Garnier, "What kind of style is this. It is not a style! It is neither Greek, nor Louis XIV, nor even Louis XV!" Garnier responded, as reported by his widow, Louise, "Those styles have had their day. It is Napoleon III, yet you complain." Today, it might be something post-modern.

---

86 Kirkland, op. cit., 133.
87 Jones, op. cit., 365.
88 David Harvey, *Paris, Capital of Modernity* (New York: Routledge, 2003), 115, 212.

The building has several names: The Paris Opera, The Opera, The Salle des Capuchins, and, the name that became commonly used, The Palais Garnier. It opened in 1875 and was the most expensive building of the Second Empire. It was at the time one of the largest theaters in Europe, having 1979 seats, though it feels much smaller than that to any member of the audience.

The Palais Garnier had two features which helped to shape the experience of its patrons. A great staircase, the *Grand Escalier*, very wide, was the entry point for everyone who attended. The entrance was grand, a kind of theater in itself. In addition, the public walking and lounging areas in the building outside the theater, on every level, were very wide, again affording an opportunity to see and be seen, to converse, to play the *flâneur* in the theater.

The Palais Garnier has been rightly called

> the new cathedral of bourgeois Paris... The glittering centerpiece of the new Paris... was meant to be much more than a theater in the ordinary sense. For Charles Garnier... it was a setting for a ritual in which the spectators were also actors, participants in the rite of social encounter, seeing and being seen... What Garnier is telling us is that the Opera had absorbed much of the role formerly assumed by the Church and the court in their ritualistic and social functions. The bourgeoisie, not the clergy or aristocracy, was now the dominant class."[89]

Needless to say, the area surrounding the building was redone to give the Palais Garnier the space it needed to become a kind of public monument. The neighborhood itself became haute bourgeois, elegant.

It was theater as theater, something appropriate to the new Paris. It is generally agreed that Paris is among the most beautiful and enticing cities in the world, mainly as a result of its transformation from 1853 through the Second Empire and beyond as the Third Republic continued the work with the establishment of many new squares and gardens, a reconstruction of the Hotel de Ville (1873–1892) and the Eiffel Tower, celebrating the hundredth anniversary of the French Revolution, in 1889. Modern Paris is one of several examples in modernity of the attempt to make the city itself a work of art.

---

[89] Marvin Trachtenberg and Isabel Hyman, *Architecture, from prehistory to post-modernism* (Upper Saddle River, New Jersey: Prentice Hall, 2003), 431.

Barcelona

Barcelona began the nineteenth century as did Manchester. It ended the century much closer to Paris.

In his *Handbook for Travellers in Spain* (1845), Richard Ford stated, "Catalonia is the Lancashire of Spain and Barcelona is its Manchester."[90]

Barcelona had a port and in the eighteenth century it developed into an important center of trans-Atlantic trade with the many colonies of Spain. In the nineteenth century it industrialized early, and its cotton trade, though not as large as that of Manchester, was the basis of its major export, as textile manufacturing grew. In 1833, Barcelona had its first steam-powered factory, though workers rebelled and burnt it down in 1835, again recalling Manchester's difficult early industrial years. Spain's first railway was the route from Barcelona to Materó, 30 kilometers northeast of the city, opened in 1848. Soon, a large network shaped in a figure eight centered in Barcelona served the area.[91]

The town, however, had walls going back to the late medieval period and its tradition as a military city. There were even laws which prohibited building outside the walls, on the large plains outside the town. Hence, with the beginnings of industrialization, factories and housing for laborers were limited to the already crowded area inside the walls. Moreover, jobs in industry drew an influx of new people. With the building of factories, the increased demand for housing, shops to serve the rising population, and places to provide for social needs, the city grew up instead of out. As land ceased to be available, a city which had had buildings of one to three storeys soon had as many as five to seven levels.[92]

In 1700 Barcelona had a population of 64,000. By 1859 the city had over 150,000 people living in much the same area. The population density was 850 inhabitants per hectare, among the highest in Europe. London at the time had a density of 100 per hectare and Paris had 300. As in Manchester, there were epidemics, including a Yellow Fever epidemic in 1821 that killed an estimated 20,000 people, and many kinds of health hazards. A cholera pandemic which reached Spain in 1854 killed over 236,000 in the country, more than 5,500 in Barcelona. Many people now suggested tearing down the walls so that the city could expand, breathe, and develop in a healthier manner.

---

[90] Richard Ford, *Handbook for Travellers in Spain* (London, 1845), 454.
[91] Joan Busquets, *Barcelona: the urban evolution of a compact city* (Cambridge: Harvard University School of Design, 2014), 114.
[92] Albert Serratosa, "The Eixample (Ensanche) of Barcelona (1859 and After): Theoretical and Practical Paradigm..." in Christian Hermansen, Cordua, op, cit., 212.

One of the first of these voices was Pedro Felipe Monlau (1808–1871), a doctor and early student of the public health implications of industrializing cities. In 1841, responding to a competition organized by the city council for the best response to the question "What benefits would accrue to Barcelona, and especially its industrialization, from the demolition of the walls that surround the city?," he published a manifesto entitled *Abajo las murallas! (Down with the Walls!).*

Monlau argued that to "enclose a population in strong walls is to oppose its physical development and its progress in all areas; it is to imprison its inhabitants." Six years later he published his *Elementos de Higiene Público* (*Elements of Public Hygiene*) in which he set out the importance of public hygiene and the responsibility of government to ensure it: "From the state of hygiene of a people we can determine its degree of security, liberty and comfort, as well as from the peace, liberty and wellbeing of a people one can very easily determine the hygienic conditions to which it is exposed." Large cities needed careful planning, he wrote, because otherwise they were horrific:

> The air is infected, the water polluted, the soil exhausted for great distances; life there is necessarily short; the joys of abundance are little known and the horrors of need and misery extreme. There one finds a permanent source of nervous diseases and epidemics; the asylum of crime and vice. Depravity is always in direct proportion to the spread of these enormous and dismal heaps of men; and the passions and vices they generate degrade them physically as well as morally; wounding their health as well as their heart.[93]

Barcelona's city council sent repeated requests to the national authority to have the walls torn down. Finally, in 1854 it received authorization to proceed. This decision only raised more questions. What would the expansion look like? Who would design it? What principles should guide the new development?

Ildefons Cerdà took up the task of designing the expansion simply as part of his passion for understanding and responding to the new era he defined as having begun with industrialization and locomotive transport. He had a desire to do more than build or to be a powerful and authoritarian civil servant. He wanted to develop a theory of urban studies, a multidisciplinary task.

He went to Paris. He praised Napoleon III for his "heroic valor" in his "daring reform." However, he found the French experience one which was not transferable anywhere else. First, it was authoritarian:

> We found nothing there but the omnipotent will of one man who says: Let this be done!, and it was done. At the sound of his voice, the public treasury opens up, the Municipality

---

[93] Pedro Felipe Monlau, *Elementos de hygiene publico* (Madrid, 1862), 3, 84. Translated by Adrian Shubert.

of Paris contracts massive loans, and the economic question is not of the slightest importance; thus [urban planning principle] could not be studied there.[94]

His aim was to establish principles that could be applied not only in Barcelona but elsewhere, in order to have a program of reform. In France and elsewhere in Spain "the method followed in them is not the offspring of any system nor of any general thought, for which reason it sets no precedent. It was rather a means chosen at random in order to escape, no matter how, from a pressing situation imposed by the need to do something as a courtesy to civilisation and the manner of being of modern peoples."[95]

Moreover, not only was there "a lack of systematic or general thinking in Paris," there was also what he termed "a lack of equity." Here, there is ideology, which needs to be part of reform, claimed Cerdà. He saw the removal of many from their dwellings as a result of expropriation to be the opposite of reform. He especially criticized the French experience for not having provided living quarters for those whose homes were destroyed. This, he believed, caused serious harm to many. As well, he realized the number of dwellings in central Paris would be fewer after Haussmann's work. As a result "...it is clear that... either many families must be left in the streets, unable to find housing, or else, faced with the harsh bind of having to wander with all their goods and chattels on their backs and sleep in the open, they will have no option but to live in the company of other families, paying dearly to be badly housed."[96] Cerdà also believed that the state should provide housing for all classes in a manner which would enable them to live together and in a rough equality, something that did not happen in Paris. For him, all citizens need to be considered in urban design.

Cerdà first did a topographic survey to get to know the area and what was actual. He then, in 1856, wrote his *Monograph on the Working Class*. Albert Serrasota lists three important discoveries at the time in the work. Life expectancy for the bourgeoisie is 37 years; for the working class it is 17. No housing in the current Barcelona, including that of the wealthy, has enough air through ventilation or natural light. Disease breeds in all of them. And, even more radical, the working class pay between 20–30 per cent more per square meter for their housing than the bourgeoisie.[97]

---

94 Cerdà, *Five Bases...*, op. cit., 362.
95 Ibid.
96 Cerdà, op. cit., 360–62.
97 Serrasota, op. cit., 215.

Cerdà knew the work of many socialists, but he was not a follower of any particular one. Rather, he had a sense of fairness and a belief in the equality of all citizens. These principles deeply influenced his theoretical considerations.

Cerdà was commissioned by the Civil Governor of Barcelona, the representative of the national government, to draw up a plan for the new city. He finished in 1859, with a plan entitled *Plano de los Alrededores de la Ciudad de Barcelona y Proyecto de su Reforma y Ensanche* (*Scheme of the Outskirts of the City of Barcelona and Plan for its Reform and Expansion*).[98] The Barcelona city council, which the national body ignored, was angry. It called for a new competition and in late 1859 approved a project designed by a municipal architect. However, because not only Barcelona was involved in the project—it included other jurisdictions outside of what was then Barcelona—the central government claimed the authority to choose. In May 1860 a royal decree approved Cerdà's plan.

The objections to Cerdà were several. First, in Barcelona there were attacks on the central government interfering with the will of the people as represented by its elected town council, a forerunner of what would occur in battles between Catalonia and the Spanish central government to the present day. Architects in Barcelona protested that such an important project should not be led by an engineer, for engineers were associated with technology, not beauty. Even further, socially and politically, architects were on the side of the traditional aristocratic landowners, whereas engineers were seen to be sympathetic to the new manufacturers, the industrial bourgeoisie. Cerdà was viewed as not having proper respect for the traditions of Barcelona aesthetics.[99]

Cerdà's plan proposed a grid system for the areas to be newly developed, sometimes called the *Ensanche*, sometimes the *Eixample* (both translated as the Expansion). Streets would extend from the old city, which was only 192 hectares, to the new development, which added 1,969 hectares to the city. The grid was to be composed of squares, rather than rectangles as in other cities, in order to more easily provide services and regulate traffic. Moreover, the square grids would have housing available to both the wealthy and the laboring classes. Services, carefully organized, would be equally available to all. He proposed that each of the squares have a limited number of residences taking up about half the space. The rest would be given over to common green spaces, internal paths, and a proper circulation of air for hygienic reasons. In his idealism he wanted to be certain to end the association of land use and value with economic

---

98 MaryDena Apodaca-Cahalane, *Ildefons Cerdà: The True Founder of Comprehensive Planning*, http://blousteinreview.rutgers.edu/ildefons-Cerdà-the-true-founder-of-comprehensive-planning/, 5.

99 Apodaca-Cahalane, op. cit., 4.

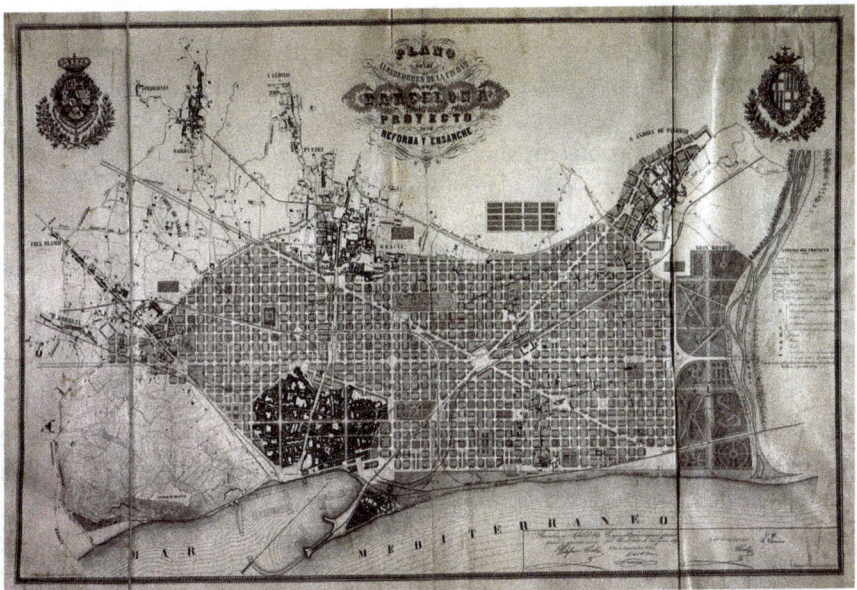

Fig. 14: Ildefonso Cerda, Map of the project of expansion of the city of Barcelona, (Cerdà's 1859 plan)

and social class. Here it is clear he was influenced by such early socialists as Cabet, Owen, and Fourier. Height restrictions, as in Paris, were planned.

The square grid system, Cerdà believed, also meant that not only rural areas could be urbanized:

> Until now, enlightenment and civilisation have displayed a resolute tendency to *urbanize* the countryside. The time has come to think about *ruralising* the major cities. This ruralisation, with the moral and material perfection of civilised man... shall easily be achieved by the adoption of the *square grid* system."

Cerdà defended his plan:

> [O]ne specific effect of its regularity, branded as a grave defect by some common and superficial minds, is to allow for the town to be sited in the manner most consistent with precepts of hygiene in relation to the influence of the sun and of the healthiest winds.[100]

---

[100] Cerdà, op. cit., 129.

In order to facilitate travel, Cerdà included a number of diagonal roads. Hence, in place of the usual ring roads and radial streets, there was a different order. Streets were wide, there were main streets, diagonals, and two main avenues leading to the port of Barcelona. Cerdà expected that equal spaces and attention would be given to locomotion and pedestrians.

The grid system was planned for an efficient allocation of goods and services. For the population at the time, roughly 250,000 in the old town and the *Eixample*, Cerdà proposed "33 schools, three hospitals located on the edge of the city for hygienic conditions, eight parks, 10 markets and 12 administration buildings." His design was also egalitarian in giving all the inhabitants equal access to services.[101]

Unique to the plan was the provision of chamfers, cut corners, on street corners and buildings in the *Eixample*. Here, Cerdà returned to the initial experience which gave him an insight into a new era dawning. The chamfers resulted in a more spacious perspective and a far more pleasant pedestrian area. Moreover, it enabled the horse-drawn vehicles of the time to turn far more easily, giving greater visibility. Cerdà went further and argued that what he called "the advent of locomotive urbanization" meant the Barcelona would be prepared for whatever would result from trains and other vehicles. He even predicted the automobile in his 1867 work:

> The experience of the immense benefits that the new locomotion brings with it cannot be more favourable... So even today, when it might be said that the new system is in its infancy, there may not be found a single urban man who would not wish to see the locomotive functioning inside the urbs, along all its streets, in front of his house, in order to have it constantly at his disposal... [T]he electric telegraph has already become widely popularised... Similarly, we also believe that after a series of experiments, trials, and evolutionary advances, the locomotive will ultimately be urbanized.[102]

A number of commentators have remarked on how Cerdà's concerns are also those of the major social theorists of the time, including Marx and many other socialists and writers. His city was not only to be beautiful, hygienic, and green; it was also to be a place where issues of class and the distribution of goods and services were to be addressed and ameliorated.

He did not ignore the matter of how to deal with the question of the sanctity of private property. He argued that "respect for property should not be an obstacle to urban improvements." In the end, private property, for Cerdà, "must be

---

[101] Apodaca-Cahalane, op. cit., 6.
[102] *General Theory of Urbanization* (1867), 728.

sacrificed to the general interest, welfare, and prosperity." The state has an obligation to what Rousseau earlier called the *volonté générale* (general will), what Cerdà termed the requirements of the "public good."[103]

He was no anarchist, and never went so far as to invoke Proudhon's famous 1840 proposition that "property is theft." Rather, he went back to the classical world, citing the Roman formula for the defence of the state: *Salus populi suprema lex* (The good health of the people is the highest law).

> ...[A]ny enlightened and paternal government which understands the exalted significance of its mission must subordinate and sacrifice private interest to this perfecting, these positive aims of moral or material welfare, neutralising... the intrigues and the hypocritical scheming which private interest is wont to resort to with considerable dexterity."[104]

Cerdà, consciously or not, was a follower of Saint-Simon, though very different from Haussmann. All this was to be done democratically and fairly, Cerdà implied in other writings, not arbitrarily as in Paris. But in the end, the general will is the guideline.

Cerdà's *General Theory of Urbanization* was planned to be four volumes. Only the first two were published in his lifetime, the first dealing with urbanization in general, the second based on data from his work on understanding Barcelona.[105] It is the first work on urban development that tried to make urban studies into what was called in Europe a "science" (Spanish: *ciencia*, French: *science*, German: *wissenschaft*), meaning a discipline based on clear rules of knowledge. In the case of some new disciplines of the nineteenth century, such as history and, later, sociology and anthropology, it was also to be founded on empirical data. Hence, the work was the first of its kind in urban studies.[106]

Volume I, the theory, is a doorstop of a book, worthy of comparison with those of other system-makers of the time, including Comte, Spencer, Marx, and Darwin. It includes material on the history of cities, on different kinds of cities, and many particulars, including the idea of the house, on various roads, and on urban functions.

Locomotion is important to Cerdà and he identifies and extensively discusses four kinds as they developed: pedestrian, equestrian, dragging, and wheeled.

---

103 Cerdà, op. cit., 366.
104 Cerdà, op. cit., 366–67.
105 *General Theory*, 24–25, intro. by Vicente Guallart.
106 Several commentators have remarked on Cerdà using concepts important in the development of ideas in the mid-nineteenth century, including some used by Marx and Darwin. This will be commented upon later in this book.

The last analytic chapter is entitled, "On the Introduction of Wheeled Locomotion into the Urbs."

He introduces a kind of dialectic, though he never uses the term:

> The important question is whether, at a time when such a deep and radical transformation is taking place, this monumental product of successive eras (none of which were similar to ours) can adapt, accommodate and adjust to the new needs that we are already seeing today (and more are emerging every day), which were never expected or even dreamed of in earlier times.[107]

Cerdà cautions that the changes cannot be resisted. History for him, as Marx stated, is a "stern taskmaster." Nor ought humans to simply destroy all that is present in order to build some utopian new civilization, as Bazarov suggested as the only option for Russia. There is what he called "a middle ground." He admits that the middle ground he advocates is itself radical, for it will make great changes in how people live and relate to one another. However, if done properly, "the aim is to free [humanity] of the ills from which they are suffering and to provide [humanity] with the legitimate advantages of which they are currently deprived."[108]

We are, he asserted, social beings and there is an "instinct of sociability" in our nature which created urbanization. We can progress—he means this morally as well as objectively—through study and application. Don't blame previous generations for not doing what we should be doing today. "Men of the age of steam and electricity!," appeals Cerdà, "Why do you not say: we are a new generation, we have new means and resources, powerful, irresistible, incomparable to those of previous generations; we lead a new life, with new ways; old urbs are nothing but obstacles in our way... Why do you not exclaim *recedant vetera: nova sint omnia?*" (Let the old depart, let everything be new)

Cerdà cautions, as do others of his time and generation, that we cannot be weighed down by the past. The example he uses is how many are responding to the new railways and locomotives in copying the ring road of their ancestors, copying what was done to respond to another era, that of coaches and carriages. Like many of his time, the past is to be understood. But the new era of Becoming means it also must be transcended.

Cerdà died nine years later, in 1876, long before he could have seen the results of his plan for Barcelona. The *Eixample* developed very slowly. Only 100 hectares were built in the first 20 years after the approval of his plan. Main ave-

---

107 *Theory*, 56.
108 *Theory*, 57.

nues were created even later. The idea of building houses only on two sides of the squares of the grid was abandoned in favor of a greater density than Cerdà had desired. It is less green than his proposals. Moreover, even his grid system and proposal for equality of services could not fully withstand the capitalist development of some of the land. The area around the main avenue, the Passeig de Gràcia, became the most desirable and valuable location. In some cases, though not with skyscrapers, height regulations were not followed.[109]

However, Cerdà's grid layout with wide streets, his innovations of chamfers and the planting of trees, as well as subsequent building, did create an extraordinarily interesting and vital urban area, which transformed Barcelona into one of the models of urban development, especially for cities of its size.

Barcelona had the double good fortune of having extensive land outside its walls which could be developed and the plan of Cerdà. In the latter part of the nineteenth century Barcelona became the home of a number of creative people, including talented architects, who were the leaders in shaping a new style called Catalan *Modernisme*.

This *Modernista* era was guided by two aesthetic concepts. First, there was the influence of Art Nouveau, which was then in many cities in western Europe. Art Nouveau was both architecture and design. In Barcelona at the time industries related to construction were established as a result of the needs in the *Eixample*. Hence, there were tapestry and furniture design centers, tile establishments, and various ornamental workshops, all employing many craftspeople in addition to doing industrial design.

The second idea was a national one in the Catalan area. There was an attempt to identify and support Catalan identity and culture, not only in design and architecture, but as well in language, literature, and the visual arts. Spain was (and is) a federal grouping of a number of distinct regions and cultures. Catalonia was not much different then than Scotland, the Basques in both Spain and France, the Welsh in Great Britain, or the Walloons in Belgium, all desiring not only some political autonomy but a revival of their national identity. Hence, a good deal of Catalan Gothic was adapted by the new architects to the style of Art Nouveau.

The beginning of the architectural part of the movement was the Café-Restaurant designed by Lluís Domènech I Montaner for the 1888 Barcelona Universal Exposition. As in Britain and elsewhere, arts and crafts were encouraged, though in Spain there was no difficulty in combining craftsmanship with industrial manufacturing. Many creative architects flourished in addition to Domè-

---

[109] https://historyofbarcelona.weebly.com/plan-Cerdà.html.

nech, including Enric Sagnier, Josep Puig I Cadaflach, and, best known now, Antonio Gaudi. They and many others built much in the *Eixample* and contributed to its being Catalan in style and culture.

The single building emblematic of the new *Modernisme* and Catalan style (comparable in importance to the Palais Garnier in Paris) is in the old town, the Palau de la Musica Catalana, built between 1905–08, designed by Domènech.[110]

Manchester humanized itself in the second half of the nineteenth century and forward. It lost some of its centrality as well when the industrial revolution moved from textiles to chemistry and other ways of creating wealth. It is a kind of cautionary tale on how the singular pursuit of wealth can also create misery, and on the need to respond humanely and positively.

Paris remains a world capital, though France's place in the world—as well as that of England—is not as central as it was in the nineteenth century. Its inner core, inside what is now the Boulevard Périphérique, remains a contender for being the most beautiful city on the planet. Its outer edges need lots of attention.

Barcelona developed differently, for it tore down the walls and expanded on land 10 times its original size. As well, it adopted a grid pattern which has become a model for others of similar size in Spain and elsewhere. The result is a city which is perhaps the most interesting "second city" (Madrid being first in Spain) in Europe.

All three cities tell much about what was happening in Europe in the middle of the nineteenth century. The idea of "the city" did not merely grow. It was reshaped into something that could reflect modernity and deal with the modern experience.

---

[110] *Barcelona and Modernity*, ed. William Robinson et. al. (New Haven: Yale University Press, 2007), 144–50, 158–61, passim.

Chapter VII
Karl Marx and Friedrich Engels: Understanding Industrial Society

In 1859 Karl Marx was living in London with his family, an exile, often in financial difficulty. Friedrich Engels was in Manchester, working in his family firm. The two men began their intimate and productive friendship and collaboration in 1844, when they met for the second time after Engels finished his *Condition of the Working Class in England*. They saw one another regularly when in the same city and corresponded nearly every day when not. Engels often contributed money to help the Marx family, Karl, his wife Jenny, and their three daughters. The family was poor, and their clothes and possessions were sometimes in the pawn shop. Karl and Jenny Marx had seven children, but they suffered the loss of four of them in childhood, including the tragic death of their beloved son Edgar, at age six, in 1855.

There was in 1859 no such thing as Marxism, nor were there Marxists. Marx was then well known in socialist and communist circles and well regarded as an intellect and philosopher by many. However, at that time there were many socialists and many socialisms. Proudhon was perhaps the most revered. Bakunin was a rival. Lassalle would soon be popular. All would change with the publication of the first volume of *Das Capital* in 1867, and with the ideas and events emanating from the International Workingmen's Association (IWA), usually called the First International (1864–1876). By 1876 Marx was famous and notorious on the European continent, his ideas in the forefront of socialist thought.

In the 1850s Marx spent much of his time in the British Museum, researching and writing. He had some income as a regular European correspondent for the *New York Tribune*, a newspaper founded by Horace Greeley, which grew in that decade to become the largest daily in New York City.

In England, he was largely unknown and mostly ignored. Unlike Mazzini and Kossuth, who were popular and lionized, Marx was regarded as just another person unwelcome in his native land and several other places, who found refuge in London. As long as he obeyed the law and stayed out of trouble, he was tolerated. Somehow, London and England could house many radicals without suffering any serious problems.[111]

---

[111] See Sabine Freitag, ed., *Exiles from European Revolutions: Refugees in Mid-Victorian England* (New York: Berghahn Books, 2003).

In 1857 and 1858 Marx's thoughts and research were compiled in notebooks never published in his lifetime. They first became public in 1939–41, published in Moscow under the title *Foundations of the Critique of Political Economy*, and commonly referred to as the *Grundrisse*, the German word for foundations. He did publish a work in 1859, *A Contribution to the Critique of Political Economy*. Both the *Grundrisse* and the *Contribution*, which did not sell well, are preliminary to *Das Capital*. Much was rewritten from both and then inserted into the masterful 1867 work.

The most important part of the *Contribution* today, scholars agree, is the preface, in which Marx tells the reader, in roughly five pages, the history of the development of his materialist theory of history.

He studied law, he reported, but he "pursued it as subject subordinated to philosophy and history." He then became a journalist and in late 1842 became the editor of a Cologne daily newspaper, the *Rheinische Zeitung*. His editorial work led him to a consideration of questions related to the economy, about which he knew little at the time.

As a result, Marx looked to Hegel's work on the philosophy of law. Marx's findings became central to everything in his theories:

> My inquiry led me to the conclusion that neither legal relations nor political forms could be comprehended whether by themselves or on the basis of a so-called general development of the human mind, but that on the contrary they originate in the material conditions of life, the totality of which Hegel... embraces with the term 'civil society'; that the anatomy of this civil society, however, has to be sought in political economy.[112]

People enter into relations of production as they lead a social existence, relations connected to how the material forces of production are used. These relations are economic and social, out of which comes the "legal and political superstructure," which is part of social consciousness: "It is not the consciousness of men that determines their existence, but their social existence that determines their consciousness." It should be noted that the issue of what "determines" means in this context has been interpreted differently by followers and commentators. Some take it almost fundamentally. Others, including the early Marx, see it as what we might call conditioning rather than determining.

At some point, Marx continues, the productive forces in a society come into conflict with the relations of production. The result is revolution leading to changes in the superstructure. A new consciousness arises out of the contradic-

---

[112] Karl Marx, *A Contribution to the Critique of Political Economy* (New York: International Publishers, 1970), 20.

tions in material life. And it is orderly: "No social order is ever destroyed before all the productive forces for which it is sufficient have been developed and new superior relations of production never replace older ones before the material conditions for their existence have matured within the framework of the old society." Hence, humans resolve to solve social problems when material conditions are ripe for them to be able to do so.

He identifies four different modes of production in human history—Asiatic, ancient, feudal, and "modern bourgeois." He asserts that the bourgeois mode is the one that will produce the last antagonisms, the last set of contradictions. He suggests that all that has occurred is what he calls "the pre-history of human society." It is thus a given, though he does not say it here, that a new era awaits.

Marx then discusses his collaboration with Engels beginning in 1844 and the *Manifesto* (1848), "jointly written," as an important moment, in addition to other works. The last decade, he notes, has been spent studying political economy—learning what needs to be learned to develop a clear and coherent theory and ideology—and with his journalism. His purpose in writing the preface, he states, is to show that his views "are the outcome of conscientious research carried on over many years." He is correct. In order for Marx to develop his complex and comprehensive system he needed to learn a good deal more than can be taught in any single discipline. He mastered philosophy, economics, law, and politics in developing what he thought of as his science of society.

Just as Engels' 1844 *The Condition of the Working Class in England* contains the basis of much that will come after that time, so too do the *Economic and Philosophic Manuscripts of 1844* by Marx, written before his second meeting with Engels and the formation of their relationship. Intended to have been a book, unfinished and not published until many years later, it reflects much that will be developed by Marx.

The two works are similar in tone and in moral concerns. They display an unusual meeting of two fine minds that will continue the rest of their lives. Reading these two works and the *Manifesto*, it can be argued that no others at the time understood the effects of the industrial revolution on its first generation better than did Marx and Engels. They provided important new concepts and categories.

Early in the *Manuscripts* Marx reflects on what he calls "estranged labor" in the new industrial society. He notes that he wishes to explain the relationship between private property, greed, and labor, among other matters. He insists that he is not building theory on any assumptions based on abstractions, but rather on economic data.

This is the beginning of Marx's idea of developing what he would term historical materialism. Hegel, the greatest philosopher of the previous generation, was an idealist in that he believed ideas propelled the development of history. Marx studied Hegel and appropriated, as we shall see, his idea of the dialectic as a way of explaining historical change. However, the famous *bon mot* was that Marx "stood Hegel on his head," substituting material conditions as the important base of analysis. "In direct contrast to German philosophy which descends from heaven to earth, here we ascend from earth to heaven," he wrote in *The German Ideology* (1845–46).[113]

In the new industrial society, Marx first argues, the worker actually becomes poorer the more he produces. This is because what he produces is what is valued, a thing. And "the *increasing value* of the world of things proceeds in direct proportion the *devaluation* of the world of men." Labor now not only produces a commodity. The laborer is himself treated as a commodity.

The result of industrial work is that which is produced by the laborer is both alien to the producer and has a power independent of him. Both the product and the worker are objectified. Hence, the appropriation of his labor for the worker is source of estrangement, what Marx terms *Entäusserung*, alienation.

This feeling of alienation is an important feature of industrial work and is important in revolutionary thought. The concept of alienation in the Christian era was associated with the idea of feeling estranged or alienated from God. Marx here and elsewhere secularizes it, and relates it to our daily activities and material conditions, something many other philosophers will do after him.

Marx also indirectly deals with the issue discussed by Mill, the need to be able to be free to shape the plan of our lives, for humans to be progressive beings. If we are treated as objects, as producers who are commodified, it also means that our labor is not meaningful, that it exists outside of us and has a power over us. We are no longer subjects, but objects. We become things rather than persons.

Further, Marx argues, since the worker is hired to produce something that belongs to someone else, his work is an "alien activity not belonging to him." The result thus goes beyond being alien to the thing produced. In addition, there is estrangement from the self because his own activity "is turned against him, neither depends nor belongs to him."

We are, claims Marx, species beings by nature. We are social beings and are part of a larger community. The nature of industrial work also causes "the estrangement of man from man." He elaborates on this in other works, as will the industrial novelists in their stories about the industrial revolution. Class is pitted against class and often worker is pitted against worker in the quest for

---

[113] Tucker, op. cit., 154.

finding work that will give one subsistence. Competition is part of the marketplace. Some win and some fall.

Alienated labor and private property are associated together. At one point in the *Manuscripts* Marx rhetorically claims that private property is the result of estranged labor, something difficult to comprehend both historically and conceptually. He then moves away from causation and talks about them being reciprocal.

Marx, and Engels as well in his writings, attempts to redefine what it means to be human and how to live in a humane manner. If in the Enlightenment the formula was that of Descartes, *Cogito ergo sum* (I think, therefore I am), for Marx and most early socialists (the terms socialism and communism were often used interchangeably in the 1840s and 1850s) it was *Homo Faber*, humans as makers, shapers, creators.

We shape ourselves: "[F]or the socialist man the *entire so-called history of the world* is nothing but the begetting of man through human labor, nothing but the coming-to-be of nature for man..." We make history and we do so through our labor, which is how we express our nature. If our labor is appropriated by someone else, then we cannot be free. It is not only slavery that takes away our humanity. In industrial society it is also the way labor is organized. Communism is thus "the next stage of historical development in the process of human emancipation and recovery." It will enable humans to freely shape themselves. If private property is separated from its power of estrangement, it can then provide what is essential for all humans, and our activity will provide enjoyment as well as meaningful activity. There is no question that the early Marx and Engels are both concerned not only with our social and political conditions, but also with our psychological well-being.

Marx goes on to discuss "the power of money in bourgeois society." Money is both power and property. In a society in which everything is commodified, money is the means of maintaining life. Marx uses material from both Goethe's *Faust* and Shakespeare's *Timon of Athens* to illustrate this matter. Money trumps all. As Marx states, money means that one's authority has little to do with one's personal qualities. He gives several examples, including the idea that if one is stupid and has money, he simply buys talented people. As he puts it, "he who can buy bravery is brave, though a coward." In our age when money buys honour, this should be easily understood.

Money (and property as money) now is associated with one's worth as a person. Money as a way of dealing with desire and human relations hence is itself alienating. Later, Marx and Engels will discuss the commodification of life in bourgeois society as something less than human.

At the close of this discussion the young Marx muses that in truly human relationships you don't commodify. In a human world "you can only exchange love only for love, trust for trust, etc." That which we do for free tells much more about us than that which we do for money. Marx and Engels want what we would call authentic human relationships.

Another early unpublished set of reflections was what has come to be known as the "Theses on Feuerbach" (1845). They are a small set of 11 notes made by Marx which were found by Engels after Marx died and were published in 1888.

Ludwig Feuerbach's best-known work was *The Essence of Christianity* (1845), written at a time when a number of theologians were reflecting and revising traditional views about the idea of a creator and his relation to his creations. Feuerbach proposed what he termed "an anthropological essence of religion." Instead of God creating culture, God is understood as a creation of culture. Our concept of God is a projection of human needs and human understanding. Religion is one of the ways we humans deal with the sufferings in our lives.

Marx suggested that Feuerbach still did not pay enough attention to the material side of life, that he remained an idealist even in his revisions of Hegel. Hence, as Marx states in the seventh thesis, "Feuerbach...does not see that the 'religious sentiment' is itself a social product, and that the abstract individual whom he analyses belongs in reality to a particular form of society."

The most famous of the theses, often quoted, is the last, the eleventh: "The philosophers have only *interpreted* the world, in various ways; the point, however, is to *change* it." This idea of the role of thought and the relationship between thought and action is something relatively new and very much part of modernity.

Philosophy, for Marx and others in the 1840s and 1850s, is no longer an armchair activity. It is designed to both interpret the world, that is to give us tools to understand what has happened and is happening, and then to transform a world of Becoming into something better. Indeed, philosophy in this sense may not merely be thought. By acting, one is making certain choices about what is, what ought to be, and how to get there, and between means and ends. Action on the barricades or in a revolution or in the streets is thought. The two cannot be separated.

This position is something that might well have guided the revolutionaries of the late eighteenth century. It certainly motivated the socialists (and many liberals) in the 1840s, leading to the revolutionary year of 1848.

It is also something that had a bit earlier been taken up by one of the thinkers most admired by Marx, Johann Wolfgang von Goethe (1749–1832), in his great masterwork, *Faust* (1832). Goethe saw a revision of the Faust legend as an exemplar for his times. His Faust, a philosopher bored and depressed in his study, does not find life interesting or meaningful. He is visited by Mephistopheles who tells him to get out of his laboratory and into the world. Be an agent of change, he tells him, be someone who brings the benefits of modernity to humankind. Faust acts. In his actions he becomes, as Marshall Berman put it, a developer.[114]

This Faust not only changes himself; he changes the world. He constructs and builds. And he also destroys, tears down homes, and appropriates nature, all in the name of progress. In effect, though neither Cerdà nor Haussmann ever cited Faust, he does what Cerdà accused Haussmann of doing, and what would also be done in Barcelona by Cerdà's successors. Now, according to Marx, the philosopher must get his hands dirty. He must be, if necessary, a revolutionary in order to transform our material conditions for the better.

In 1847 the Communist League, to which both men belonged, met in London to hold its second Congress. At its end, Marx and Engels were asked to write a Manifesto which would contain the beliefs of the organization. After several drafts, Marx wrote the final version, published in February 1848 as the *Manifesto of the Communist Party*, certainly one of the most influential documents of modern times.[115]

Marx was certainly the senior. John Kenneth Galbraith made an appropriate summary: "Engels always considered himself a junior partner, and so, without doubt, he was. But that does not lessen his role. Had he not been the junior partner, much for which his senior partner is known would not have been done."[116]

The opening words of the *Manifesto* are well known: "A spectre is haunting Europe." Marx was right and wrong. He called it the spectre of Communism of

---

[114] Berman, op. cit., ch. 1.

[115] The collaboration between the two men was one that lasted the remainder of their lives, nearly two score years. It is not possible to clearly decide on their individual contributions. However, Engels always deferred to Marx as the senior partner and more profound intellect. In 1886, in a footnote near the end of his book, *Ludwig Feuerbach and the End of Classical Philosophy*, Engels wrote: "Here I may be permitted to make a personal explanation. Lately repeated reference has been made to my share in this theory, and so I can hardly avoid saying a few words here to settle this point. I cannot deny that both before and during my 40 years' collaboration with Marx I had a certain independent share in laying the foundation of the theory, and more particularly in its elaboration. But the greater part of its leading basic principles, especially in the realm of economics and history, and, above all, their final trenchant formulation, belong to Marx. What I contributed—at any rate with the exception of my work in a few special fields—Marx could very well have done without me. What Marx accomplished I would not have achieved. Marx stood higher, saw further, and took a wider and quicker view than all the rest of us. Marx was a genius; we others were at best talented. Without him the theory would not be by far what it is today. It therefore rightly bears his name" (https://www.marxists.org/archive/marx/works/1886/ludwig-feuerbach/ch04.htm).

[116] John Kenneth Galbraith, *The Age of Uncertainty* (Boston: Houghton Mifflin, 1977), 89.

course, given the origin of the *Manifesto*. It was larger than that. It was the spectre of revolution, made in the main by other socialists and radicals, which would start in France (at the end of February) and make its way through much of the European continent, 1848 being, until 1989–91, the most revolutionary year in Europe's history.

But the words signal something else for the work of these two young communists, 29 and 27. The *Manifesto* is not only ideology, it is rhetoric. It was written in both an accessible and elegant style, using all of the tools of language to make its case, including sentences and phrases that have made their way onto ordinary discourse and became instantly recognizable. In places, it is eloquent.

Its first section, titled "Bourgeois and Proletarians," goes to the heart of the argument. It starts with a bold claim, a one-sentence paragraph: "The history of all hitherto existing society is the history of class struggles."

The statement is first what is sometimes called a philosophy of history. How does change occur in human history? What are the motivating forces? These questions were asked by historians from the time of Herodotus and Thucydides in the Greek world and especially in the early nineteenth century as history became a discipline. The answers were varied. They included the idea that "great men," the Newtons and Napoleons, appeared and all was transformed. Some looked to politics and the state as the agent of change. A few, St. Simon among them, suggested economics need be taken into consideration. And Hegel, the giant philosopher in the generation before Marx and Engels, discussed the importance of ideas.

Not so for the authors of the *Manifesto*. Change occurs first on the level of class. That is the major category of analyses of the past and the present. Moreover, it occurs as a result of "struggle," a concept enormously important in the period which included Marx and Darwin. One of Marx's favourite words in all of his writings was "*kampf*," struggle.

The struggle occurred between "freeman and slave, patrician and plebeian, lord and serf, master and journeyman," between classes, as they say, "oppressor and oppressed...in constant opposition to one another." It ends in either a new society or in the ruin of both classes in contention.

Something else is hovering about this formulation. It is the dialectic, favored by Hegel and adapted by Marx. Hegel was a dialectical idealist; Marx and Engels were, as they often said, dialectical materialists.

The dialectic is a kind of logic different from traditional formulations. Normally, when A and B are in opposition to one another and compete, the result is understood as one of them triumphing over the other. That's good Aristotelian logic.

Chapter VII Karl Marx and Friedrich Engels: Understanding Industrial Society — 151

In the dialectic, the understanding is more complex when dealing with social and political affairs, or in Hegel's case with ideas. In the dialectic, if A and B are opposed, they contend and the result is C. And C will over the course of time produce its opposite, D. And change goes on.

The beauty of the dialectic is that it not only accounts for change in history but is also part of a process in the present, as Marx and Engels recognised. We can see in Burckhardt's study of the Renaissance an example used in the *Manifesto*. The struggle between lord and serf in Italy in the early Renaissance resulted in the victory of neither. Rather, there appeared the new urban dweller, the new personality of the Renaissance, and the new despots. As Marx and Engels put it, "The modern bourgeois society… sprouted from the ruins of feudal society." If we understand the French revolution of 1789 as a bourgeois victory, as did Marx and Engels, then it should be remembered that among the first acts of the revolutionary national assembly was the law to end feudalism; soon the Assembly also approved the Declaration of the Rights of Man and of the Citizen.

The claim has another element. Classes not only tend to develop in opposition to one another; they also reflect a relationship between "oppressed and oppressor," as we saw in Engels' study of Manchester in the early industrial revolution. Hence, the *Manifesto* tells us that "the modern bourgeois society has not done away with class antagonisms. It has but established new classes, new conditions of oppression, new forms of struggle in place of the old ones."[117]

Marx and Engels put forward a philosophy of history and a theory of social change. But there is also a moral element. The terms "oppressed" and "oppressor" are laden with values, in addition to being a claim about what is real. The Communists will claim not only to be correct about the pace and meaning of the past. They will also claim to be on the side of the oppressed, to be morally right. The *Manifesto*, and the theory, is both normative and valuative.

Therefore, claim the authors, at this moment we are in "the epoch of the bourgeoisie," an epoch in which there are new class conflicts and struggles. However, now class conflict is "simplified." Modern society "is more and more splitting up into two great hostile camps, into two great classes directly facing each other: Bourgeois and Proletariat."

This split in society is the result of discovery, trade, new markets and the new manufacturing system. The feudal system could no longer function. Now there is steam, machinery, and a market that takes in the world rather than lo-

---

[117] Many have taken the idea of struggle to heart in claiming their rights. For example, in his *Letter from Birmingham Jail*, Martin Luther King wrote, "We know through painful experience that freedom is never voluntarily given by the oppressor; it must be demanded by the oppressed."

calities. The result is that the classes that existed in the middle ages are no longer at the center of commerce or production. The guilds are marginalized in favor of the industrial middle class, the new millionaires, the "modern bourgeoisie," which "is itself the product of a long course of development, of a series of revolutions in the modes of production and exchange."

Moreover, this new bourgeoisie has, in the age of industry, become the dominant political class and power. It has taken over the modern state. For Marx and Engels, political power cannot be understood outside of considerations of political economy. The state is not at all neutral, and never has been. It reflects the dominant class. Hence, "the executive of the modern state is but a committee for managing the common affairs of the whole bourgeoisie."

They then (remember it is 1848) discuss the bourgeoisie in an unusual manner. The argument is that the bourgeoisie is a revolutionary class, one which might pretend to be conservative, but has radically changed the way we live and relate, produce and consume, socialize and organize. They overthrow a whole social order. Indeed, the next several pages praise the bourgeoisie in terms like those of a publicist. They have, argue Marx and Engels, accomplished much. However, as history will develop, they will be overthrown, for "what the bourgeoisie...produces, above all, is its own grave-diggers," the proletariat. The dialectic of history is, for Marx and Engels, relentless.

The accomplishments of the bourgeoisie are many and they are major. They ended the "motley feudal ties" of the serf-lord relationship. Now, they suggest, all is "naked self-interest," commodification. The bourgeoisie constantly transform the nature of production, nothing remains the same. Then, poetically: "all that is solid melts into air, all that is holy is profaned, and man is at last compelled to face with sober senses, the real conditions of life, and his relations with his kind."

The new bourgeoisie has revolutionized communication and transportation, and they have made production and consumption cosmopolitan. As Europe and the West develop, other areas are compelled to imitate them, to adopt bourgeois modes of production and society. Moreover, Marx and Engels above all recognize the new modes of living. Towns and cities have eclipsed the countryside. Speaking as the moderns that they are: "[The bourgeoisie} has created enormous cities, has greatly increased the urban population...and has thus rescued a considerable part of the population from the idiocy of rural life." People are now living in crowded towns, production has been centralized, politics are now national, not local.

They tell the reader that the bourgeoisie, in its short period of existence and then dominance, "has created more massive and more colossal productive forces than have all preceding generations together." Nature has never been more in the power of human beings than with the new machinery, the railways and telegraph, chemical agriculture, and the building of canals and dams. The new creative energy, it is suggested, far surpasses other wonders, including the pyramids, the Roman aqueducts, and the Gothic cathedrals. This, in an age where there are yet no skyscrapers and where the modern "cathedrals," the railway terminals, are just in process. It is a magnificent achievement, this transformation and the world of Becoming.

Marx and Engels discuss other matters, too, that relate to people's daily lives. The halo image is used, as noted earlier in this study. All occupations that had been esteemed are, in the new world of money, "stripped of [their] halo." Now all is commodified. The priests, lawyers, doctors, scientists, and poets are now like the rest of us: "wage-laborers." Communists would see this as a good thing, because they would like to eliminate class and class conflict by making everyone a laborer. In addition, they view the family through a political economy lens: Its "sentimental veil" is now gone, as the new order "has reduced the family relation to a mere money relation."

Hence, where we are now is the result of changed productive production and exchange forces being no longer compatible with the old feudal social and property relations. The old were "burst asunder" by the new: "Into their place stepped free competition, accompanied by a social and political constitution...and by the economic and political sway of the bourgeois class."

In another powerful image, bourgeois society is likened to a sorcerer, to someone who has magical powers. And, of course, the new inventions and productivity seem like magic compared to rural and feudal society. However, it is "like the sorcerer who is no longer able to control the powers of the nether world whom he has called up by his spells." The new industrial world is out of the control of the men who own it because it has its own logic, which includes economic crises, the need to always expand markets, and always to grow. Hence, this society is also "forging the weapons" of its own destruction. As well, it has created the proletarians who will do the job.

Marx and Engels then discuss matters that each dealt with separately in 1844—the nature of labor in the factory system, the exploitation of workers, and the fact that labor has "lost...all charm for the workman."

The proletariat struggle to find a way out of their dilemma. They destroy machines in an attempt to turn back development. But in doing so they mistake their enemies. They get concentrated in factories and cities, learn how to act collectively, form unions, and take collective action that is sometimes violent. The two men are clear that in the end the bourgeoisie, which grows smaller in number, will not yield. The proletarians gain consciousness and recognize that the system will not substantively change. They will get concessions, but not freedom.

As Marx and Engels put it, for the proletarian, "Law, morality, religion, are to him just so many bourgeois prejudices." As they say later in the document, "The ruling ideas of each age have ever been the ideas of its ruling class."

The above assertion relates to a distinction made by both men in many of their works. They are materialists, and they believe that what is important occurs in what is sometimes termed the substructure, the way goods and service are produced and distributed, the conditions of labor and life. As well, each society has a superstructure—its ideology, its law, its religion, its myths, and its beliefs. The superstructure, they claim, is the way the ruling class justifies its power and advances its interests. It is not neutral, it is part of domination. Hence, for example, laws forbidding certain people to vote are not about justice, only about power. If religion is, as Marx earlier famously put it, "the opium of the masses," it is an opium which contributes to the power of the bourgeoisie while at the same time providing consolation to the powerless. For them, that in some new world the poor shall inherit the earth is a convenient fiction supported by those who now own it.

At an appropriate moment, when the classes are in clear opposition, the proletariat acts. Not only do the bourgeoisie create their own grave-diggers: "[Their] fall and victory of the proletariat are equally inevitable."

The second section of the *Manifesto* is titled "Proletarians and Communists." Marx and Engels tie the two together. Their aim at the time is the same: the "formation of the proletariat into a class, overthrow of the bourgeois supremacy, [and] conquest of political power by the proletariat."

The largest issue proposed by the Communists is then directly addressed: the abolition of private property. They make it clear that they are not talking about ending the property gained by individuals as a result of their labor. Rather, they refer to "modern bourgeois private property," by which they mean the accumulation of capital, which is used to exploit individuals who have no capital. Capital is not merely wealth. They call it a "collective product," and claim that it not only has economic power, it also has great "social power." Hence, capital, not personal property, should be "common property." If so, it immediately loses its class nature (technically, though it has never happened, the revolution can be made without violence. The proletariat can appropriate the productive forces and its property. The owner no longer owns the factory, etc. If he were someone like Engels, whose family owned the factory, he might be asked to continue to manage the factory and be paid a reasonable wage. He then ceases to be a bourgeois and joins in unity with the proletariat).

The authors even address the reader, as powerfully as Charlotte Brontë did in *Jane Eyre*, published only a few months before the *Manifesto*. "You are horrified," they say, "at our intending to do away with private property." And they add that in bourgeois society "it is already done away with for nine-tenths of the population." No, they say, you worry we want to do away with your property. Yes, that's what we hope to do.

At some point the proletariat will gain political power, take capital from the bourgeoisie, and centralize production in a state guided by their own class. This upheaval might have to take place by fiat, what they call "despotic inroads on the rights of property, and on the conditions of bourgeois production," but it might be necessary in order to "revolutionize the mode of production." They don't say so here, but they and others have said it often in discussing conditions in the mid-nineteenth century. The bourgeoisie will be upset and claim the proletarians are violating the law and the sanctity of private property. For the Communists, the law is made by the bourgeoisie and private property is a means of doing violence, even a kind of enslavement, to most others. As they say, "Political power, properly so called, is merely the organised power of one class for oppressing another."

The Communists have a brief program, something most socialist groups put forth to succinctly explain and clarify their ideas. It was radical at the time, though most of it seems tame now. It includes the abolition of landed property and inheritance, a progressive income tax, the establishment of a national bank to centralize credit matters, putting the means of transport and communication into the power of the new state, the development of both factories and rural areas according to a plan, "equal liability of all to labor," a better organization of the relationship between cities and rural areas, "free education for all children in public schools," and the ending of child labor in factories as it existed at the time.

The envisioned result in replacing the old bourgeois society will mean the end of the old classes and the old enmity, a new fairer society in which there is a shared humanity: "[T]he free development of each is the condition for the free development of all." Marx and Engels disliked what they called "utopian socialism." However, they do have something of a utopian ideal in mind in what they regard as the necessary development of the next stage of history.

The next short section of the *Manifesto* discusses other socialist and communist literature and movements. In all, the authors cite 11 other types of socialist movements, indicating something that will bedevil European socialism for the next century and more. Although the word "socialism" was relatively new, various ideologies developed as alternatives to liberal and conservative ideas and positions. On the left, many who called themselves socialists immediately spawned a variety of doctrines and beliefs. In this section Marx and Engels attack the other positions vehemently, something socialists of all kinds will do henceforth, sometimes reserving their greatest contempt for those closest to them on the political and social spectrum. This kind of criticism is what Freud called, in another context, the "narcissism of minor differences," finding your greatest enemy in that which is closest to you, as was the case in the development of nationalism in Europe as well. Socialists seem to have loved baiting other socialists. Indeed, near the end of the nineteenth century, the term "revisionism" was used by those who regarded themselves as true Marxists as an insult to others who "deviated" from the new gospel. Some versions of socialism and communism became dogma and woe unto those who differ even a little from the true believers.

Among the various other socialisms and socialists mentioned in this section are Proudhon, who is termed a representative of bourgeois socialism, something that fools the proletariat into thinking well of the bourgeoisie, and German socialism, which they claim "belong(s) to the domain of... foul and enervating literature." The trio of St. Simon, Fourier, and Owen, a generation previous to the *Manifesto,* are grouped as utopian socialists, even though they have different programs. Marx and Engels see them as opposing political action by the proletariat, erroneously thinking that they can arrive at a better society through reform and developing new modes of living. For the communists, this stance makes them enemies of the working class.

The *Manifesto* ends with a coda of less than two pages. Marx and Engels announce that the communists will be pragmatic and support different groups in different countries in order to work to overthrow the existing society. However, they will continue to advocate and teach the recognition of the dichotomy between the bourgeois class and the proletariat. In the context of that moment, the two have the communists look to Germany, which they believe is about to have a bourgeois revolution with a proletariat more advanced than in England and France. Moreover, the bourgeois revolution in Germany will be "but the prelude to an immediately following proletarian revolution."

And then the last grand rhetorical flourish, very powerful: "Let the ruling classes tremble at a Communistic revolution. The proletarians have nothing to lose but their chains. They have a world to win. Workingmen of all Countries, unite!"[118]

The excellent and insightful Isaiah Berlin was correct when he said of the *Manifesto:* "No summary can convey the quality of its opening and closing pages. As an instrument of destructive propaganda it has no equal...." He also

---

[118] Many readers will know that the reference to losing their chains is a nod to Jean-Jacques Rousseau (1712–1778) who in his *The Social Contract* (1762) opened Chapter I with the famous declaration: "Man is born free, and everywhere he is in chains."

stated that its influence on future generations is without parallel outside of religious history.[119]

However, we might note that the notion of what is a sacred text has also changed in modernity, and there are other political documents which have gained a similar status in the West, including the English *Magna Carta*, the American *Declaration of Independence*, and the French *Declaration of the Rights of Man and of the Citizen*. We not only have the invention of new traditions in the West, we have the apotheosis of individuals and their works, the three most clear in the twentieth century being Gandhi, Mandela, and King.

Still, as outstanding political and social rhetoric, there is little to match the *Manifesto* in its time. There are some passages in Mill's *On Liberty* which qualify, though not the whole work. The closest from someone who is a politician are some of the speeches of the American Abraham Lincoln, especially his *Gettysburg Address* and *Second Inaugural*.

1848 was a kind of test for revolutionaries and socialists. Revolution began in France in February and spread to much of the rest of Europe. There were hopes for a new order on the continent. But it turned out to be, as the English historian G. M. Trevelyan said, "the turning point at which modern history failed to turn." Reaction won the day by the early 1850s. Where socialists expected to gain, they were sometimes shot, as on the streets of Paris in the "June Days," June 23–26, 1848. Where nationalists hoped to gain, they were defeated, as in the domains of the Austrian Empire and on the Italian peninsula. Many went to prison; many became exiles.

Marx was among those whose hopes were destroyed. However, he responded with analyses of the situation in several ways, including a deep and complex work regarded as one of his finest studies, *The Eighteenth Brumaire of Louis Bonaparte*, published in 1852.

The reference in the title is to the coup d'état that Louis Napoleon Bonaparte successfully made in France on the night of December 1–2, 1851, giving him the opportunity to remain president beyond one term and soon greatly increasing the powers of the executive. His uncle, Napoleon Bonaparte, executed his first coup on the Eighteenth Brumaire of the new calendar, introduced in France for a time after the first French revolution, November 9, 1799, in which he overthrew the existing government of the Directory and became First Consul with great executive power. The work was written and published before Louis Bona-

---

119 Isaiah Berlin, *Karl Marx: His Life and Environment*, 2nd ed. (London: Oxford University Press, 1948), 161–62.

parte would further emulate his uncle and become Napoleon III in December 1852.

In the preface to the second edition of 1869, Marx stated that his purpose in writing the work was to "demonstrate how the class struggle in France created circumstances and relationships that made it possible for a grotesque mediocrity to play a hero's part." How could it happen that a revolution begun to reform a monarchy and possibly introduce a republic could result in the authoritarian regime of the coup of 1851?

Marx opens the study with some reflections on history and society. The first two sentences have resonated and been used ever since by many. "Hegel remarks somewhere," mused Marx, "that all great, world-historical facts and personages occur, as it were, twice. He has forgotten to add: the first time as tragedy, the second as farce."

He sets the stage. The tragedy for Marx is the French revolution of 1789. The farce (also a tragedy, but not played out as such) is 1848–1851. 1789 was the victory of the bourgeoisie, which turned into a tragedy for the lower classes. 1851 will also result in a victory for the wealthy, but it turned into something that lacked the depth and hopeful developments of the events of 1789.

In the second paragraph Marx then reflects on freedom and necessity: "Men make their own history, but they do not make it just as they please.…" We are, in our material life, part of a context, and that context affects our choices and possibilities. This thought is reminiscent of the statement made later by the Spanish philosopher José Ortega y Gasset about his identity: "*Yo soy yo y mis circumstancias*—I am myself and my circumstances." Hence, what is possible is limited by where and when we happen to be. We are free to choose, but we are also proscribed by context. We cannot make the industrial revolution in the Renaissance of Burckhardt.

Even more, "the tradition of all the dead generations weighs like a nightmare on the brain of the living." We sometimes have models from the past which limit our perception of what is possible. We adapt past models to understand what might be new and thereby misunderstand what is occurring. He claims that "the Revolution of 1848 knew nothing better than to parody…1789 and the revolutionary tradition of 1793 to 1795." The nightmare, as he puts it about a page later, is that "from 1848 to 1851 only the ghost of the old revolution walked." However, we are in a new era. If there is to be a real social revolution in the nineteenth century it "cannot draw its poetry from the past, but only from the future." It must deal with the reality on the ground, the substructure of life.

Marx then moves from philosophy to the material world, going deeply into the society, politics, and events of the time in order to analyze what happened and why. In doing so, however, there are a number of important concepts and ideas discussed which help to guide his narrative and help us to understand his ideas.

He distinguishes between bourgeois and proletarian revolutions in terms of how they proceed. The bourgeois revolutions, especially those of the eighteenth century, go quickly and seem successful early in their development. However, he suggests, "they are short lived." Proletarian revolutions are more chaotic, move back and forth, and sometimes seem to succeed only to find that their opponents gain strength again. At some point, after much effort, they might find success.

Hence, he points out early in his analysis, an astute observer might have seen "that a terrible fiasco was in store for the revolution." In the early days, in the new government after the February revolution, many groups participated, including those who were royalist of one sort or another,[120] bourgeois republicans, members of the petit bourgeois class, and representatives of socialist and democratic workers. Soon, in May, when the National Assembly met as a result of elections, it became apparent that the bourgeoisie were in control. Hence, the proletariat attempted to dissolve it and were defeated both in the assembly and on the street in the June Days, when there was a proletarian uprising in which 10,000 people were either killed or injured.

France had been a bourgeois monarchy from 1830 to 1848. In the course of the development of society, claims Marx, it "can only be followed by a bourgeois republic." Faithful to his earlier works and analyses, Marx comments that a bourgeois republic is simply a means of a single class dominating all other classes. Politics is a function of social structure and power. This revolution (and others) is to be understood by an analysis of class interests and conflict. Moreover, those who opposed the proletariat demonized them, claiming that they, the bourgeoisie, represented stability, property, family, and tradition against those who were dangerous socialists.

Marx also included some important reflections on the nature of class as he understood it. In discussing the peasants and their positions at the time, he notes that while they are many, "a vast mass," and live in conditions which are similar to one another, they do have many relations reciprocally: "Their mode of production isolated them from one another, instead of bringing them in to mutual intercourse." The small- holding peasant and his family are nearly self-sufficient. Villages are small, made up of what he terms a few score of families separated from one another. And then a bunch of villages become a *département*. As he puts it for the reader, "the great mass of the French nation is formed

---

[120] There were three royalist traditions by then in France, Bourbons (reigning until 1830), Orleanists (1830–1848), and Bonapartists (1804–1814).

by simple addition of homologous magnitudes, much as potatoes in a sack form a sack of potatoes."

Then, he defines what he means by class:

> In so far as millions of families live under economic conditions of existence that divide their mode of life, their interests and their culture from those of the other classes, and put them in hostile contrast to the latter, they form a class. In so far as there is merely a local interconnection among these small peasants, and the identity of their interests begets no unity, no national union and no political organization, they do not form a class.

The result is that these isolated peasants are not capable of acting in their class interest. They are represented by an authority, often acting against their interests. In the case of France at the time, as a result of the coup of Louis Napoleon, it is the central executive that takes on this role. Bonaparte gets the support of the conservative peasant, not those in favor of revolution. The small-holding peasant often supports the old order, seeking to protect what he has. He does not have the consciousness of being a member of a class, nor does he have the ability to see beyond his narrow interests: "It represents not the enlightenment, but the superstition of the peasant; not his judgment, but his prejudice, not his future, but his past."

Thus, classness is related to consciousness and the ability to transcend one's own existence and see beyond one's limited life. You can be a laborer and not be a member of the class of laborers unless you are conscious of yourself as part of the larger society and a member of a group. The conditions of labor in industrial society help to create the class of laborers, the proletariat. However, the peasants at the time do not have conditions which encourage consciousness of oneself as a member of a class. This explains their support of the conservative order: "Bonaparte...represents not the revolutionary, but the conservative peasant."

The coalition that Bonaparte relies on to keep his power is thus one comprised of the bourgeoisie, the small peasants, the army, a variety of monarchists, and the support of the Church. It should be noted that in much of the scholarship around the variety of revolutions and changes in government in France from 1789 to 1871, there is general agreement that a vast political change could not be successful without the support of the bourgeoisie. Hence, the "success" of the revolutions of 1789 and 1830, and the "failure" of 1848. With the fall of Napoleon III during the Franco-Prussian War in 1870 there came into being the new Third Republic. Adolphe Thiers, the first president of the Republic, famously stated as the new state was forming amid great differences of opinion: "The Republic is the form of government that divides us the least." It, too, had the support of a majority of the bourgeois class.

In thus explaining the success of Bonaparte in being elected president in 1848 and his ability to carry out a coup in 1851, Marx claimed that he was the best alternative for the crucial support of the middle class: "Manifestly, the bourgeoisie had now no choice but to elect Bonaparte." Then, in parentheses, he cannot help making an insightful and somewhat insulting remark: "Despotism or anarchy. Naturally, it [the bourgeoisie] voted for despotism."

Two main things occur with Bonaparte in power. The centralization of the French state continues, thereby preparing the way for an eventual takeover by the proletariat when the time is appropriate. Both Marx and Engels were consistent in saying one cannot hasten history or skip over necessary developments. The proletarian revolution will occur when the dialectic between the classes is ready for it. Marx is saying that the attempt at revolution came too soon.

Secondly, he predicts rightly that the government of Bonaparte will sponsor industry and trade. The state will be active in many areas, including public works and banking. Like the redoing of Paris, it will be a government highly acceptable to the bourgeois class. The argument is one in accord with the Marxist emphasis on economics and the social substructure. Bonaparte will gain political power with the aid and support of the bourgeois class which sees him as supporting their economic interests. They don't need to have political power as long as they have a government supporting their economic authority.

He adds one other group to those that supported Bonaparte. It is described by a term invented earlier by Marx and Engels, the *lumpenproletariat*. The word (*lumpen* is German for rag) refers to those members of the lower class who, like some rural peasants, do not have class consciousness. Often they are referred to as beggars, vagrants, prostitutes, petty thieves, and some unemployed. Marx was especially contemptuous of this group of people who were reactionary in their politics and supportive of a politician who he saw as an enemy of the proletariat.

Bonaparte, Marx claims, tries to be a "patriarchal benefactor of all classes." But he is not. He insultingly calls Bonaparte the "chief of the Society of December 10" in terms of how he keeps power by buying support. That so-called "Society" existed in late 1848 and 1849. It was headed by supporters of Bonaparte. They recruited members of the lumpenproletariat to cheer at public appearances of Bonaparte, and to use violence if necessary against those opposed to him.

Marx closes with two of his favourite images. Bonaparte, he says, tries to be a "conjurer." However, he will fail because of the contradictory demands of his supporters. Moreover, he tarnishes the state, "he divests the whole state machine of its halo, profanes it and makes it at once loathsome and ridiculous." He, too, like his uncle, will fall.

It is a very sophisticated analysis, done before Louis Napoleon Bonaparte became Napoleon III. It is also instructive, though Marx might disagree and say

Fig. 15: Honoré Daumier, *Members of the Charitable Society of December 10 Demonstrating their Philanthropic Activities*, October 1850

that in doing so one is falsely using the past to understand a different period with different material conditions, as he cautioned in the opening of the essay.

The models of Napoleon I and Napoleon III will reappear, with force, in the first half of the twentieth century. They pulled self-coups, as it were, coups on their own governments and states and moved to authoritarian positions in the name of the people. They claimed authority from the people directly rather than any legal status or constitution. They governed by decree and manipulation. And they talk about the greatness of their country. Moreover, they too had the support of the capitalists and the reactionaries who believed, rightly or wrongly, that they could control the despot.

Hence, the coming to power of the two Napoleons can be regarded as providing a template for fascists later on in the period after the First World War. On the right, in Italy, Germany, Portugal, and Spain, similar coups occurred. Poland had a military dictatorship in 1926 and Yugoslavia ended parliamentary democracy in 1929. Strong fascist groups existed in the 1930s in France, Belgium, the Netherlands, Greece, and Romania. Often, fascist leaders had their own para-military groups in support, similar to Louis Bonaparte. It would not be appropriate, totally out of context, to call the two imperial Napoleons fascists, but it is sometimes tempting.

The summing up of the views of Marx and Engels regarding their ideas on socialism was left to Engels. In 1880 he published *Socialism: Utopian and Scientific*, which became a popular document and a handy statement about Marx and Marxism in a number of languages. It also contained a clear discussion of where they thought the proletarian revolution was heading.

It is in three sections, sometimes summarizing what the two men had said in other writings. The first is on the Enlightenment and the utopian socialists. Engels praises Enlightenment thinkers for their contributions, especially their scepticism and reliance on reason. However, "we know today that this kingdom of reason was nothing more than the idealised kingdom of the bourgeoisie." The proletariat were hardly considered. The "utopian socialists," St. Simon, Fourier, and Owen, are discussed in greater detail than in the *Manifesto*. The judgement is the same—they were limited in seeking reform rather than revolution. Their socialism was not a sure scientific understanding of material conditions.

A second section is given over to Hegel's ideas, their contributions and their limitations. The introduction of the dialectic as a mode of understanding is given its due. Further, "In this system [that of Hegel]—and herein is its great merit—for the first time the whole world, natural, historical, intellectual, is represented as process, i.e., as in constant motion, change, transformation, development..." Now we need to see and comprehend the reality of Becoming.

However, as he and Marx said elsewhere (and often), Engels judges Hegel as limited—by his time and especially by his idealism. Hence, Hegelianism was a "colossal miscarriage" which needed to be corrected. The problems of Hegelian thought (not including the dialectic) brought about the need for a new examination of history and for the development of a science of society based on facts, rooted in materialism. Adopting a rhetorical technique favored by Marx, the inversion of a statement to sharpen the contrast, Engels stated, "now a materialistic treatment of history was propounded, and a method of explaining man's 'knowing' by his 'being,' instead of, as heretofore, his 'being' by his 'knowing.'"

The third part summarizes the two men's materialist ideas and the process of moving from the medieval period to what is called socialized production in the industrial age. Engels traces the growth of industry, the bourgeoisie, and the proletariat, and discusses the contradictions that appear. Finally, *"The contradiction between socialised production and capitalistic appropriation manifested itself as the antagonism of proletariat and bourgeoisie"* (his italics). He quotes Marx from *Das Capital*, likening the laborer to a Prometheus bound to a system which produces misery and slavery. Again, the rhetoric is both normative and valuative, in their terms "scientific" and moral.

The capitalist society has periodic economic and social crises built into it which cause a centralization of industry and the establishment of joint stock companies as well as state ownership of property and some production. But remember, he notes, the modern state is the state of the capitalists, part of the power and organization of bourgeois society. As all this develops it moves towards the moment when the small capitalist class is in dialectical opposition to the vast number of proletarians in the population. Moreover, once the proletarians understand the nature of the productive forces in capitalist society, these forces "can, in the hands of the producers working together, be transformed from master demons into willing servants...*The proletariat seizes political power and turns the means of production into state property*" (his italics).

Now, in several pages, we have, near the end of the lives of both Marx and Engels, a statement about the future tailored to the understanding of people who might also read the *Manifesto*.

By taking power and changing the ownership of the means of production, the proletariat, as Engels states, abolishes itself as a class because it changes the nature of society. And the state, which had been part of the way the bourgeoisie remained in control and exploited others, is also seen as no longer necessary: "It dies out."[121] Society is no longer separated into two classes, one the oppressor and the other oppressed. Engels again emphasizes that this is all a process. It cannot happen until the development of production reaches a sophisticated point. It "presupposes a degree of historical evolution." As they both often said, context matters. At some time, the means of production in the hands of one class, as well as its political control and its ideology, "become not only superfluous but economically, politically, intellectually, a hindrance to development."

And then, dramatically, writing this in 1878, Engels purposely composes a short sentence: "That point is now reached." It was not reached in 1848 or even in 1870, with the Commune of Paris. It is here. The time has come when "the socialized appropriation of the means of production" is necessary not only to do away with the extravagance and waste of bourgeois culture and society, but also because it will harness greater productive forces than can be done under the bourgeois order.

The passages read as a special moment in the whole history of humanity.

> The possibility of securing for every member of society, by means of socialised production, an existence not only fully sufficient materially, and becoming day by day more full, but an

---

[121] Sometimes that last phrase is translated as "It withers away."

existence guaranteeing to all the free development of their physical and mental faculties—this possibility is now for the first time here, but *it is here.*

Engels uses an image that he and Marx had evoked earlier. When this special moment occurs, humans move from "animal conditions of existence into really human ones." What is meant is that heretofore most humans had to always concern themselves with obtaining the essentials needed for survival—food, shelter, etc. Now humans can face nature without being dominated by it; now humans will not be dominated by a social and political system which is oppressive and out of their control. Humans can now shape their "social organization" and their "history." Man will "make his own history... It is the ascent of man from the kingdom of necessity to the kingdom of freedom."

There is an unspoken assumption behind this conclusion, one central to the thought of many in the mid- to late-nineteenth century. Once socialized production becomes the norm in place of capitalism, not only will the proletariat work out the new system, but production will continue to increase the amount of goods and services. In short, there will be abundance, and that abundance will end the unfair scarcity that was part of the capitalist system because the system of production and distribution will be transformed. Hence, individuals will know that their basic needs will be taken care of, and they can now shape themselves (not unlike the goal of Mill). The result is that "Man, at last the master of his own form of social organization, becomes at the same time the lord over Nature, his own master—free." Scientific socialism is the theory; proletarian action is the means.[122]

Engels' and Marx's goals have been seen as noble ones by many since their time. However, as we know, their vision of a fairer and finer future has not come into being. In some places, western and northern Europe especially, social democracy grew into something that changed the conditions of capitalism, as discussed by many, including the influential work of Eduard Bernstein, *Evolutionary Socialism* (1899). In other places in Europe, where the claim is that Marxism was being put in practice, it became an authoritarian system that would repel both Marx and Engels if they had seen it. It is Lenin who invented the single-party state and Stalin who established the gulags.

What Marx and Engels do offer, however, is both profound and meaningful. Their understanding of the first part of the industrial revolution, through which

---

[122] Engels' formulation would be later used by others, many with no relationship to Marxism. For example, when Virginia Woolf in 1928 asked what a woman needed to be free she answered, "money and a room of one's own." Many who attacked colonialism asked for the appropriation of the economic goods of their territory.

they lived, is the best template at the time. From their moment on, no serious historian or sociologist could analyze a society without taking into account the economic means of production and distribution.

The categories and modes of analyses of the two men contributed to a much deeper sense of what was happening. Terms such as "substructure" and "superstructure," "material conditions," "relations of production," and others came to be used by many. Their system of analysis has a language and internal logic that is enormously seductive. One must be careful to remain critical rather than adopting it by default. Nonetheless, they made profound contributions to understanding what Baudelaire called "modernity."

As well, they appealed to many who saw the injustice of the industrial system and the injustices built into unfettered capitalism. In the end, their moral outrage was at least as important as their self-professed "scientific" analysis. John Stuart Mill was enormously sympathetic to some of the socialist program after the publication of *On Liberty*, though he notes in his *Autobiography* that he opposed what he called the "tyranny of society over the individual" of most socialist movements. Still, Mill attacked the injustices of industrial society, including the way property was distributed and the way the industrial system was organized. The fact that wealth was related to the accident of birth was deeply offensive to his sense of social justice.

Critiques and appreciations of the work of Marx and Engels fill libraries. In the context of this study, there are three matters that bear discussion. In modernity it is nationalism that persisted and became the main force behind the development of European states. Certainly socialism in many states made great headway, but the idea of the nation remains the main ontological base of the modern Western world.

Marx and Engels are part of a long list of intellectuals from St. Simon to the present who have wrongly predicted that nationalism was either a phase of modern development and/or that it will disappear. "The working men have no country," they claimed. For Marx and Engels it was class that was seen to be what would unite people in Europe. "Workers of the world, unite," they said. "[You] have a world to win." Hence, as consciousness developed, the workers of, for example, Germany would realize they have more in common with the workers of France or Britain than they do with other Germans.

That did not happen. When war seemed possible in 1914, there were many in peace movements and/or in socialist groups and parties that advocated a general strike among the workers of Europe to stop what seemed a senseless and, for many socialists and communists, a war caused by the capitalist structure. It did not come close to happening, as workers in the various countries supported their own government and their own people against others. Workers joined others in singing the then German national anthem, *Heil dir im Siegerkranz* (Hail to Thee in Victor's Crown), or *La Marseillaise* or *Land of Hope and Glory* as they marched to war. Even more dreadful, workers were among those who encouraged their children to join the military in defence of their country, and they witnessed many of them wounded or dead. Even those countries which became "communist" —be it the Soviet Union in the inter-war period, or the several eastern European satellites after the Second World War—organized and appealed to their followers in nationalist terms. Today, nationalism, in its traditional guise or via the new populism, is still on the march.

Second, there is the omission of any basic concern for human rights in Marx and Engels, even though they do want labor to be free. In their analysis of the condition of Europe of their time, human rights were viewed as a "bourgeois" concern, part of the bourgeois state and social order. And, in context, they made sense, for if you look at the areas that supported human rights, they were those who limited rights to certain people. Women had no rights anywhere. In Britain, rights were effectively reserved for the privileged, as many have pointed out. In the United States, there was slavery, and racial prejudice was legal. In France, rights did not address poverty and often were denied to those on the radical left.

However, one of the ways that many groups have made important gains in modernity is through an appeal to the kind of arguments put forth by Mill, to natural rights and the rights of the individual, to the idea of universal human rights. It is communists and fascists who deny the validity of rights and who insist that one conform to the views of a collective authority. Rights matter for many of the things that Marx and Engels stood for, including the rights of children and the dispossessed, the marginal and the weak, the oppressed and those suffering. Democratic socialists all over the continent now acknowledge this matter.

Lastly, there is the issue of human nature, as they understood it. For Marx and Engels, simply looking at what was happening, they believed that the societies of mid-nineteenth century Europe were sick. Engels, in *Socialism: Utopian and Scientific*, discusses what would happen when the proletariat took power. One assumption was that there would be abundance, that there were enough goods and services in society to look after everyone's basic needs. A second assumption was that this abundance would occur under a new classless society, that each person would appropriate according to their needs in the name of belonging to a place that took into account the greater good.

In short, if society was sick, humans did evil because they were forced to do so in a flawed society. We, as readers, understand why Hugo's Jean Valjean stole a loaf of bread—because the system deprived him of the ability to feed his family. That is, human beings are basically good, and are made evil by their material conditions. Change the material conditions and we will perforce behave decently towards our fellow human beings.

Not so. Decency and benevolence did not occur. As we have experienced the two world wars, depressions, bombs, the Bloodlands, genocide organized by states, etc., we have changed our view of human nature from the more optimistic one of the age of progress. We are closer to Yeats ("The best lack all conviction, while the worst/ Are full of passionate intensity") or Freud (*Homo homini lupis*) than we are to Condorcet, St. Simon, Marx or Mill.

As we know, those states which did become "communist" in one way or another have not institutionalized anything like equality of condition. They all produce what Milovan Djilas called *The New Class*, in his insightful work of 1957. They can be commissars who obtain dachas, good living quarters, and special food to leaders of China, who enrich themselves and whose children are known as princes and princesses. They can be part of the kleptocracy of present-day Russia or the corrupt leaders of Venezuela and other states.

Not that capitalism necessarily produces a more equal condition. The United States is these days ruled by a wealthy oligarchy that manipulates laws and power in its own interests. It has an extraordinary gap between the few rich and the many poor, and it leads the world in the percentage of people who are incarcerated. And protests against the one percent in many countries reveal many lesions in the frameworks of social democracies.

The human nature issue is central to viewing Marx and Engels in context. They thought too well of us in believing what we might establish after a proletarian revolution. There is no doubt that were they to learn some of the things that have been done in their names, they would be morally appalled. Such is what sometimes happens to the founders of new forms of thinking and living, be they religious leaders or the creators of a new faith in the future. They contributed much and they did much good in their time.

Chapter VIII
The Anti-Moderns

Modernity was not welcomed or even accepted by some in 1859. It was seen to be something to be resisted, even attacked.

The Roman Catholic Church

In 1864, the Papacy and the Roman Catholic Church issued, as an appendix to the encyclical *Quanta cura*, the *Syllabus Errorum*, the Syllabus of Errors. The eightieth and last of those listed tells what the document was actually about. It is an error, it said, to believe that "The Roman Pontiff can, and ought to, reconcile himself, and come to terms with progress, liberalism and modern civilization."

The encyclical was the result of the Roman Catholic Church's view of itself as under siege in an age of nationalism, liberalism, and belief in progress. The pope at the time, Pius IX (1846–1878), started out as a liberal but became a reactionary who wanted the Church to have the status it had had in the medieval era. He was supported by a majority of bishops and others in the hierarchy, who were Ultramontanes—those who placed strong emphasis on the power and authority of the papacy.

The Church had receded in significance and influence in modernity, starting with the French Revolution of 1789, the promulgation of *The Declaration of the Rights of Man and the Citizen*, the confiscation of Church property by civic authority, and the growth of the secular state.

In 1848, the revolutionary tide of that year moved to Rome, and Pius IX, who had condemned the uprisings in the face of the murder of his lay Prime Minister, fled Rome in November. Mazzini and Garibaldi declared a republic in Rome and for a short time governed an anti-clerical state. Pius was restored in 1850 by the intervention of French troops, but from that moment on he regarded democracy and liberalism as revolutionary doctrines.

The Pope, it must be remembered, was then a secular ruler as well as a theological one. The Papal States in the Italian peninsula covered a large area in the middle, including Rome, from sea to sea. This area too would be transformed, not in 1848, but as a result of the wars of Italian unification beginning in 1859. By 1861, most of it was lost to the new Italian state, with the exception of Rome and Lazio, now a large administrative area of Italy with Rome at its center. The Pope himself and his temporal power were dependant on French and Austrian troops.

Still, the Church hierarchy behaved as if the last several centuries hadn't occurred. They claimed full authority over secular rulers and states, and immunity from the laws of the civic authorities to which they belonged. The Ultramontanes pushed the power of the papacy in contrast to others who saw the Church as an association of bishops, with the Pope as first among equals. When, in 1870, a bishop appealed to tradition at the First Vatican Council, Pius IX replied in the manner of Louis XIV. *"Tradizione!"* he shouted, *"La tradizione son' io!"* (I am tradition!).

The Encyclical *Quanta Cura* was clear on the dogma. The Roman Catholic Church is superior to any state, and all states must obey the laws of the Church and submit to its authority. The basis of civil society is religion; democracy has nothing to do with it. The Church is the protector of family and children. Education must be under the authority of the Church. There is only one religion which is true, that is Roman Catholicism. The state must support and nourish it to the elimination of other forms of belief.

The Syllabus elaborated on these notions in 80 propositions divided into 10 subjects. It is as extreme as one can get on the authority of the Roman Church. It attacks all forms of Protestantism as well as socialism, liberalism, and democracy. National churches are not permitted and nationalism is also an evil idea. When there is conflict between the laws of the Church and the state, the Church is in the ascendancy. The Church itself is to be treated as a special body, independent of any state, responsible to its own authority. There should be no separation of church and state. Morals and philosophy in general are subject to ecclesiastical authority. Marriage is sacred and divorce is not permitted. Tolerance of other religious beliefs is incompatible with doctrine.

Quanta cura and the Syllabus were followed first by a ban in 1868, in the decree *Non Expedit*, forbidding Roman Catholics from voting or standing for office in Italian elections. Then came the First Vatican Council, assembled in Rome in late 1869. The division at the Council was still between a large Ultramontane majority of bishops and a smaller moderate and liberal one. The issue which divided them and has become most prominent was the matter of declaring the infallibility of the pope. After much discussion, the vote on the matter among the bishops was taken on July 18, 1870. Over 50 bishops who opposed the matter left Rome in order not to vote against an issue which had great support. It passed 533–2. The pope was now officially considered to be infallible when speaking *ex cathedra* on matters of faith and morals.

The Franco-Prussian War broke out the next day. The council was prorogued for a time, and it never managed to meet again. Napoleon III pulled his troops out of Rome to help defend France. In the power vacuum, the Italian state invaded and annexed the territory. Rome now became the capital of the unified Italian state. The ban on participation in Italian politics remained in force until 1919.

The Syllabus and the events of the papacy of Pius IX troubled not only nationalists, democrats, and liberals. A number of Protestant churches were offended, as were some Orthodox Catholics and other Christians. The Roman Catholic Church as it had been in the medieval era simply was irreconcilable with modernity and its developments.

Anarchism

At the opposite end of the political and social spectrum from Roman Catholicism, another response to modernity grew into being in the mid-nineteenth century: anarchism. At first, anarchism was considered a form of socialism, one of many groups vying for followers on the left. Its two most important figures at the time, rivals to Marx and other socialists, were Pierre-Joseph Proudhon (1809–1865) and Mikhail Bakunin (1814–1876).

Proudhon, a printer and a self-educated Frenchman, came into prominence in 1840 with the publication of his influential work *What is Property?: An Inquiry into the Principle of Right and Government.*

Proudhon began the work with an answer to his question, a paragraph much quoted by others who were troubled by what was happening in liberal states and in the industrial revolution:

> If I were asked to answer the following question: WHAT IS SLAVERY? and I should answer in one word, IT IS MURDER, my meaning would be understood at once. No extended argument would be required to show that the power to take from a man his thought, his will, his personality, is a power of life and death; and that to enslave a man is to kill him. Why, then, to this other question: WHAT IS PROPERTY! may I not likewise answer, IT IS ROBBERY, without the certainty of being misunderstood; the second proposition being no other than a transformation of the first?

"Property is theft," became the most used English translation, and it has resonated since in the debates about authority and the state, and on the streets of many European cities.

Proudhon understood property to mean that which is public. We all have the right to the possession of the essentials of life. We do not have the right to appropriate and use that which is public for our own ends.

The enemy of the anarchists is not simply property and those who own and use it for their own benefit. It is foremost the modern state. Anarchists argued that the state is oppressive. It is in the modern world "sovereign," meaning that it sees itself as legitimately coercing citizens on the basis of its own interests.

Anarchists acknowledge the importance of society. Indeed, Proudhon in *What is Property?* stated that human beings are social by nature. For them, society is more important and prior to the state. Some early anarchists argued that there is a need for organization, for what we might call government. However, that organization should rise naturally and spontaneously from the social circumstances and should not be the modern coercive state. For them the state has no legitimacy other than power and authority, which robs us of our agency. Even further, the state as it has come to be is unjust and it uses its authority and law to permit injustice to flourish. Even before Marx published, Proudhon argued that the modern state exploited people on behalf of its own interests and/or those of the dominant group. In 1851 he summed up his attitude:

> To be GOVERNED is to be watched, inspected, spied upon, directed, law-driven, numbered, regulated, enrolled, indoctrinated, preached at, controlled, checked, estimated, valued, censured, commanded, by creatures who have neither the right nor the wisdom nor the virtue to do so. To be GOVERNED is to be at every operation, at every transaction noted, registered, counted, taxed, stamped, measured, numbered, assessed, licensed, authorized, admonished, prevented, forbidden, reformed, corrected, punished. It is, under pretext of public utility, and in the name of the general interest, to be placed under contribution, drilled, fleeced, exploited, monopolized, extorted from, squeezed, hoaxed, robbed; then, at the slightest resistance, the first word of complaint, to be repressed, fined, vilified, harassed, hunted down, abused, clubbed, disarmed, bound, choked, imprisoned, judged, condemned, shot, deported, sacrificed, sold, betrayed; and to crown all, mocked, ridiculed, derided, outraged, dishonored. That is government; that is its justice; that is its morality.[123]

Anarchy for Proudhon is "the absence of a master, of a sovereign." In contrast, a society which recognizes the independent power of private property is arbitrary and capricious. It permits and legitimizes abuse and behaviour contrary to a fair and orderly society. It robs us of our power to shape our lives in ways appropriate to our tastes and concerns.

Proudhon also borrowed from Hegel in using the dialectic to explain where he wanted social organization to go. The thesis is communism, which he regards as a basic "expression of the social nature." The antithesis is property, the contradiction of communism. If property is "the exploitation of the weak by the strong," then communism is its negation, "the exploitation of the strong by the weak."

---

[123] *General Idea of the Revolution in the Nineteenth Century*, trans. John Beverly Robinson (London: Freedom Press, 1923), 293–294.

Like Marx and other socialists, Proudhon was fond of the rhetorical device of inversion. Marx would soon use it against Proudhon in challenging anarchism. In 1846 Proudhon published his *Philosophie de la misère (The Philosophy of Poverty)*. Marx replied in 1847 with his own book countering Proudhon, called *Misère de la philosophie (The Poverty of Philosophy)*.

For Proudhon and other anarchists, communism "is essentially opposed to the free exercise of our faculties, to our noblest desires, to our deepest feelings." It negates the individual. The synthesis to be supported is liberty. Liberty, as he understands it, only exists in a social context, and in equality. There will be order, but it is an order chosen by individuals: "As man seeks justice in equality, so society seeks order in anarchy."

Proudhon's anarchism agrees with Marx and Engels that the revolution of 1789 did not complete the task of reorganizing society. Changes in government were made, but the economic issue was basically ignored. What is needed is association, which is voluntary and local. A new revolution would end "the rule of capitalists, usurers, and governments, which the first revolution left undisturbed." Then, workers would "take over the great departments of industry which are their natural inheritance."

Opposed to the idea of the state and its government would be a new concept of contract. This is not at all the contract discussed in Rousseau's *Social Contract* (1762) or that of other liberal political philosophers. Society is primary, that is, in front of government. Free individuals will make arrangements among themselves in order to establish an appropriate and new economic order which would end the need for a coercive political organization.

> In place of laws we will put contracts; no more laws voted by the majority or even unanimously. Each citizen, each town, each industrial union will make its own laws. In place of political powers we will put economic forces... In place of standing armies we will put identity of interests. In place of political centralization, we will put economic centralization.[124]

The idea is to organize society through voluntary associations from below. The anarchist premise is that the best arrangements are those made between individuals which then provide for workers to control their lives. The assumption is that decentralization is the way to go, localisms are the key to choosing how we live our lives. Proudhon was clear that once arrangements were made, once one entered into a contract, that contract must be honored and kept. His anarchism is not that of throwing bombs or doing whatever one chooses at any given moment.

---

124 George Woodcock, *Anarchism*, (Peterborough, Canada, Broadview Press, 2004), 114.

Fig. 16: Honoré Daumier, *Proudhon*, 1850. "Apostle of socialism, enemy of capitalism and patented destroyer (*without state guarantee.*)" "At his feet rats are gnawing at his brochure." (https://gallica.bnf.fr/ark:/12148/bpt6k63555863/f66.image). A second source states: "Here we see the rats chewing on the books written by Proudhon and on [Adolphe] Thiers' oeuvre, *Du Droit de la Propriété.*"

Law will thus result from arbitration among individuals. Each workers' group, agricultural or industrial, will make its own rules, education will be organized locally by teachers and parents, and people will have control over their lives. This is a social unity, not a governmental one. Proudhon expects that in it we will find our liberty in place of the present tyranny of states and governments.

Proudhon's position on nationalism reflects his dislike of centralized authority. He came from the Jura and he associated his own identity with the locality of the Franche-Comté rather than the state of France. He wanted an end to national frontiers and was among the earliest philosophers on the left to recognize that, contrary to the vision of Mazzini, nationalism and individual liberty might be impossible to reconcile.

Whereas most nationalists and democrats supported the idea of a centralized nation-state, Proudhon saw in the many small units on the Italian peninsula the possibility of a federalist structure of autonomous regions which would liberate individuals without the need for war and without the need for a centralized government.

In 1863 Proudhon wrote *Du principe fédératif*, an attempt, not always successful, to take his ideas on contract and society and apply them to what was happening in Europe on the level of state power. Federation was the important principle. All should begin at the local level. Europe should be organized as a confederation of local groups. The nation would not be the basis of power or identity. Contracts would be made between small units. This book is a work which is not argued as well as some others of Proudhon, possibly a result of ill health. However, it is one of the few works at the time which seriously challenged the institutionalization of nationalism on the political scene.

The anarchism of Proudhon and others at the time is an attempt to cope with the new world of modernity in the mid-nineteenth century by negating centralized authority, the anonymity of being an object in a large bureaucracy, and the fear of the loss of liberty in a world out of the control of individual wills. Earlier in the century, workers responded to the industrial revolution by destroying the machines. Now, anarchists try to regain the kind of control over one's life that might have existed in the guild structures of an earlier time. Modernity—its massive size, its lack of respect for society, its support of large states—repels Proudhon morally.

Proudhon's education began when he was a young printer working for Antoine Gauthier in his native Besançon. In 1829, when Proudhon was 20, the utopian socialist Charles Fourier (1772–1837), also a native of Besançon, came to Gauthier to have his work *Le Nouveau Monde Industriel et Sociétaire* published. Proudhon was involved in printing the work and met Fourier. The two men discussed a number of social issues, which influenced Proudhon. Fourier, too, in designing his utopian community, was deeply concerned with giving individuals the opportunity to have meaningful social relations in an atmosphere of freedom. Proudhon was no utopian, but he stands tall alongside such people as Fourier and Robert Owen as one of the most influential philosophers who defended workers and their dignity.

It was the charismatic and volatile Mikhail Bakunin who turned anarchism into a movement. The Russian led a different life from Proudhon, moving from country to country, jailed for some years, always preaching revolution and transformation. He moved to Berlin in 1839 in order to study philosophy. The young writer Ivan Turgenev was then living in Berlin and became a close friend. Turgenev used Bakunin as the model for the character of Rudin, the eponymous hero of his first novel. And there is something of Bakunin in Turgenev's Bazarov, the nihilist of *Fathers and Sons*.

Bakunin met Proudhon in the early 1840s, after the latter gained fame with *What is Property?* He was influenced by Proudhon's ideas and had only praise for Proudhon's work—this from a socialist who had as deep a critical eye for those who differed from him as did Marx:

> But then came Proudhon: the son of a peasant... he equipped himself with a critical point of view, as ruthless as it was profound and penetrating, in order to destroy all (doctrinaire and bourgeois socialist) systems. Opposing liberty to authority, he boldly proclaimed himself an Anarchist by way of setting forth his ideas in contradistinction to those of the State Socialists.
>
> Proudhon's Socialism... was bound in the course of time to arrive at Federalism.[125]

However, though Bakunin agreed with Proudhon on ends, he did not agree with his mentor on the means to achieve the transformation of society. He believed along with Marx (though he disagreed with Marx on many important matters) that change could only occur through violence and revolution. In discussing his interpretation of the failures of 1848, he saw "two great questions" being posed: first, what he called the social question, the need to transform society and to focus on society ahead of government; second, the freeing of all peoples and obtaining the independence of all nations. As he put it, "The whole world understood that liberty was only a lie where the great majority...is condemned to lead a poverty-stricken existence, [and] to serve as a stepping stone for the powerful and rich."[126]

---

[125] G. P. Maximoff, *The Political Philosophy of Bakunin* (London: The Free Press, 1953), 278–79.
[126] Woodcock, op. cit., 130.

What to do? He put it in messianic terms: "We must... purify our atmosphere and transform completely the surroundings in which we live." How to do it? By "the overthrow of society." Bakunin was not the only anarchist of his time to preach violence, but he was the most influential. Thus began the anarchist move to destruction as the first means of change. Of course, Marx and Engels also stated that the revolution would not come peacefully, for the bourgeoisie will not reform or cede power. This view was also echoed in fiction by Turgenev's Bazarov, who shocks the minor aristocracy, the equivalent of the bourgeoisie, such as it was in Russia, with his belief in the need to destroy first and build later.

After six years in prison in Russia in the Peter-and-Paul Fortress from 1851–56, followed by four years of exile in Siberia, Bakunin escaped to Japan and eventually lived in Italy from 1863. He first founded a group called the Florentine Brotherhood, followed by the International Brotherhood. There was now the makings of an anarchist organization, international in scope. Bakunin claimed, "we have adherents in Sweden, Norway, Denmark, England, Belgium, France, Spain, and Italy." He also said that there were even Polish and Russian supporters, and that many supporters of Mazzini had moved to anarchism. It was a wild exaggeration, a hope rather than a fact, but nonetheless he and his movement were gaining notoriety. He was becoming a major figure in socialist circles, akin to Proudhon and Marx.

Bakunin's next battle in the struggle among socialists for ideological purity and dominance was with Marx, in the context of the politics of socialism of the International Workingmen's Association, ordinarily called the First International (1864–1876), an organization which aimed at uniting a variety of communist, socialist, anarchist, and trade union bodies into a single group working on behalf of class struggle and change.

Though there was much on which they agreed, the anarchists represented by Bakunin and the communists represented by Marx found themselves on different sides.

Agreement included the end of the revolution, one of liberty and equality in social status with the means of production in the hands of society as a whole, devoid of politics. The anarchists generally accepted the Marxian analysis of bourgeois society and its critique; the Marxists were sympathetic to the anarchist need to destroy the current social and political arrangements.

The anarchists were dialecticians, but they were not rigid materialists. Bakunin and others knew their Hegel and gave some importance to the world of ideas in helping to produce development and change. They found the Marxists too rigid in their insistence on a "science" of socialism which made the world lawful. In a letter to Marx in 1846, Proudhon wrote:

Let us seek together, if you wish, the laws of society, the manner in which these laws are realized, the process by which we shall succeed in discovering them; but, for God's sake, after having demolished all the *a priori* dogmatisms, do not let us in our turn dream of indoctrinating the people; let us not pose as the apostles of a new religion, even if it be the religion of logic, the religion of reason. Let us gather together and encourage all protests, let us brand all exclusiveness, all mysticism; let us never regard a question as exhausted, and when we have used our last argument, let us begin again, if need be, with eloquence and irony.[127]

Anarchists were very sensitive to not being bound by any system, even one that claimed to be the basis of socialism.

It was Bakunin who confronted Marx with a critique that has become famous for its insight and its prophetic wisdom. The main argument centered on the state. Bakunin wrote that he abhorred those theories, including those of both Marx and Lassalle, which argued that the workers should form a "People's State," in which the proletariat would become the ruling class. That is an error, for Bakunin claimed that any state means the domination of some people by others. Hence, a People's State "is a ridiculous contradiction." If the proletariat adopt it, they will become the very oppressors they had wanted to overthrow and eliminate. He echoed Proudhon: "a State without slavery, overt or concealed, is unthinkable—and that is why we are enemies of the state."[128]

He went further. Marxist theory, he rightly said, solves the problem of how the proletariat will govern by having a small group govern as representatives of the people as a whole. Marx, he says, claims these representatives will be workers, but he is wrong. No, said Bakunin, they will be "ex-workers, who once they become rulers... cease to be workers and begin to look down upon the toiling people."[129] They will look after their own interests, not those of others, for power does indeed corrupt. He accused Marx and others of having no understanding of human nature. He once wrote that power over others produces two things: "*contempt for the masses, and, for the man in power, an exaggerated sense of his own worth*" (his italics).[130]

The Marxist dictatorship of the proletariat is a fraud for Bakunin and the anarchists. It is a mask for a new state which will be the despotic rule of a few over the new proletariat. No one will be emancipated by the Marxian formula. The new so-called transitional state will be a simple dictatorship. No dictatorship

---

127 Proudhon to Karl Marx, https://www.marxists.org › subject › proudhon › letters.
128 Maximoff, op. cit., 286–87.
129 Maximoff, op. cit., 287.
130 David Miller, *Anarchism* (London: J.M. Dent and Sons, 1984), 88.

can have any other aim but that of self-perpetuation and the slavery of the vast majority. Freedom can only live through freedom, not through a new state or a new dictatorship. Revolution should result in the "free organization of the toiling masses from the bottom up."[131] I am a collectivist, he said, not a communist.[132]

Bakunin's critique turned out to be correct. For Marxists, he said, the state will be strengthened. The people will be "under the direct command of State engineers who will constitute the new privileged scientific-political class."[133] Any reader who has read the dystopias of Zamiatin and/or Orwell will recognize easily what Bakunin was asserting. He is the first to predict that applied Marxism would turn out to be a form of authoritarianism or totalitarianism.

Marx won the political battle for control of the First International, though it was a victory without any real gains. Socialism remained divided among its many groups and movements, each confident of their own virtue and rightness, each full of anger at other socialists who were not true believers.

Anarchism did not disappear after Bakunin. It became important in Russia in the nineteenth century and Spain and elsewhere in the twentieth century. But it essentially opposed many of the transformations of modernity and never took hold as a political alternative for any length of time. Anarchists dislike the consequences of the industrial revolution. They believe the victory of the French Revolution left the bourgeoisie in power and the laborers in their thrall. They wanted small voluntary associations, not big trade unions. The big cities were not their home. The demographic growth of the nineteenth century worked against them.

In the world after 1945, the closest thing to what the anarchists were advocating was likely the original kibbutzim in the Israeli state after 1948, though many of those disappeared as future generations made other choices. And anarchists split amongst themselves, some dropping out, others joining other left movements, some taking violence and chaos as their method. Some anarchists became good assassins, not realizing that if they killed a Czar or a political leader, another like him would immediately pop up. Their ideals are modern. Their solutions don't fit into the context of modernity. That contradiction was never overcome.

---

131 Maximoff, op. cit., 288.
132 E.H. Carr, *Michael Bakunin* (New York: Vintage Books, 1937), 356.
133 Maximoff, op. cit., 289.

Arts and Crafts; the Pre-Raphaelites

If Britain was the home of the first industrial revolution, then the Crystal Palace and the Great Exhibition of 1851 was its celebration. The Exhibition is considered the first modern world's fair, and its theme was the representation of modernity and the products of advanced industry. The Palace was massive, as befitted a building representing the modern era, three times the size of St. Paul's Cathedral. It was over 500 meters long and 39 meters high. The sheet glass, made in Britain, made possible a building with the greatest area of glass ever built, one that gave rise to much praise.

The Exhibition had over 100,000 objects and 15,000 contributors. There were exhibits devoted to various countries and eras. The main categories of exhibits were Raw Materials, Machinery, Fine Arts and Manufacturers. Catalogues were printed. Samuel Wilberforce, the Bishop of Oxford gave it his blessing. "What can be nobler than industry and work," he said,

> The exhibition, as promoting the industry of nations, is a great and noble work; it calls attention to the dignity of labor—it sets forth in its true light the dignity of the working class—and it tends to make other people feel the dignity which attaches itself to the producers of these things.[134]

However, not everyone was in awe or pleased with what they saw. A group in Britain associated with the Arts and Crafts movement believed that the exhibition showed that manufactured goods lacked beauty. Led by John Ruskin and William Morris, they attacked technology and its products as leading to a diminishment of both spiritual and material well-being. Like Thomas Carlyle, who rejected the modern age, they believed that "men are grown mechanical in head and heart."

This group began to turn back to a time when design and aesthetics were produced by individuals who had a sense of the integrated beauty of buildings, material, furniture, and fittings. Like Marx, they lamented that what was lost in the industrial revolution was the dignity of the craftsperson, the individuality and beauty of creation. Unlike Marx, they formed groups which turned back to the old ways of manufacture, arguing that beauty in daily life was as necessary as food and shelter.

They founded their own businesses, among which the most important was the work of William Morris. Their views also became part of an aesthetic on the European continent which culminated in Art Nouveau later in the century.

---

134 Gillian Naylor, *The Arts and Crafts Movement* (London Studio Vista, 1971), 26.

Individuality was valued over conformity and symmetry, craft was more important than large-scale manufacture, imitation was devalued.

This group never politicized their aesthetic to the point of arguing for a transformation of society. Rather, they were a number of small-scale producers turning out beautiful and valued objects and designs. Ironically, their work was valued by the very bourgeoisie they disliked, and was purchased by the wealthy consumers of bourgeois society.

The Arts and Crafts movement was also associated with a number of members of the Pre-Raphaelite group of painters and art critics in mid- to late-nineteenth century Britain. The name of the group indicated that they saw painting (and sometimes architecture) as most beautiful in the late medieval period, especially in the revival of the Gothic and the interest in mythical and/or religious subjects. They accepted the tradition that history painting was part of the work of artists and that mimesis was to be achieved as part of the aesthetic. They gloried in rich colours and emotional subjects. In the mid-nineteenth century, Raphael was considered the major "modern" artist of the Renaissance. Now, just as there were philosophers who looked to pre-Socratic times, there was a group of artists looking to revive what had been before Raphael.

The Pre-Raphaelites became the leading artists in Britain in the several decades from the 1850s, at the same moment the impressionists, relentlessly modern, were developing their work in France. Though rejecting modernity, those in the Arts and Crafts movement, together with the Pre-Raphaelites, made important contributions to the development of a sense of beauty in everyday life in Europe, and to a more balanced appreciation of medieval culture.

Dostoyevsky

In 1859 Fyodor Mikhailvich Dostoyevsky (1821–1881) was a Russian author and member of the intelligentsia who had had some hard years as a result of his political activities. He and other members of a group that proposed reform, known as the Petrashevsky Circle, were arrested in 1849, held in the Peter and Paul Fortress, and underwent a four-month investigation. They were sentenced to death by firing squad. In December the execution was to take place and the prisoners were lined up. At the last minute, as in a bad melodrama or a Brecht play, a stay of execution was delivered in a letter from the Tsar.

Four years of exile followed in a Siberian prison under difficult conditions. Dostoyevsky was released only to undergo compulsory military service in Semipalatinsk, a city in Kazakhstan, near the Russian border. Finally, in 1859, he was given his freedom as a result of his deteriorating health, and could return to Russia.

He wrote, among other works, *The House of the Dead* (1862) and resumed his life. He had always wanted to travel to western Europe, to see for himself what he ironically called "the land of holy wonders," a Europe which created, he said, "such a powerful, magical, alluring impression on us [Russians]" Finally in June 1862, the opportunity came and he went on a tour which included Cologne, Berlin, Dresden, Wiesbaden, Belgium, Italy, London, and Paris.

Dostoyevsky had been very sympathetic to the Slavophile position which valued Russian culture and challenged the superiority of that of France, Germany, and Britain. Now he could see for himself life in the place where "our whole life, from earliest childhood, has been geared."

Thus began one of the most creative periods in the life of this man who would soon write many works which are regarded as masterpieces, placing him among the greatest novelists of any time. They include *Notes from Underground* (1864), *Crime and Punishment* (1866), *The Idiot* (1869), and *The Brothers Karamazov* (1880).

Dostoyevsky recorded this first journey abroad in *Winter Notes on Summer Impressions* (1863), a rambling essay on the relationship of the Russian soul to Europe in all its complexity, a prelude to the first part of the fictional *Notes from Underground*. Early in the short essay, he asks "Can it be that there is in fact some chemical bond between the human spirit and its native soil, so that you cannot tear yourself away from it and, even if you do tear yourself away, you nonetheless return?"

Most of *Winter Notes* consists of meditations about London and Paris, the two most revered cities from the Russian point of view. During his eight days in London, Dostoyevsky went to the Crystal Palace and the Exposition. He tells his readers,

> You feel a terrible force....It is all so solemn, triumphant, and proud that you begin to gasp for breath. You look at these hundreds of thousands, these millions of people streaming here from all over the face of the earth—people who come with a single thought, peacefully, persistently, and silently crowding this colossal palace—and you feel that here something final has been accomplished, accomplished and brought to an end. It is a kind of biblical scene, something about Babylon, a kind of prophecy from the Apocalypse fulfilled before your very eyes. You feel that it would require a great deal of eternal spiritual resistance and denial not to succumb, not to surrender to the impression, not to bow down to fact, not to idolize Baal, that is, not to accept what is as your ideal....

He recognizes the structure as a modern site which already has a kind of holy stature, but it is an idol, and a false one at that. For, as he describes, London is a city of great contrasts, wealth and a polluted Thames, a busy life day and night and places with a "half-naked, savage, and hungry population," what he calls "a seeming disorder which in essence is bourgeois order in the highest degree." He witnesses the underside of life—masses of people getting drunk, prostitution and degradation, and a little girl who could not be more than six years old in Haymarket, bruised and already beaten by life. It is Baal and it is not to be uncritically admired.

His musings on Paris begin with a chapter titled "An Essay Concerning the Bourgeois." His conclusions are the opposite of those of Baudelaire. The bourgeoisie are depicted as selfish and greedy, valuing money above everything, willing to submit to the rule of the Emperor Napoleon III: "To amass a fortune and possess as many things as possible has become the primary code of morality, a catechism, of the Parisian."

In reflecting on the revolutionary mantra of *liberté, égalité, fraternité* he reflects that liberty is only for those with money, equality doesn't exist in bourgeois society, and brotherhood is lost in this society and in what he terms the French nature:

> What has shown up is a principle of individuality, a principle of isolation, of urgent self-preservation, self-interest, and self-determination for one's own *I*, a principle of the opposition of this *I* to all of nature and all other people....

Dostoyevsky asks the perennial Russian question, "What is to be done?" The answer is a Russian one: "Brotherhood must be created no matter what." And that brotherhood, which is possible in Russia, cannot be developed fully in bourgeois society. It is, of course, also a critique of capitalism, though the word is never mentioned.

Dostoyevsky argues that it is a higher ethic to voluntarily sacrifice oneself in the interests of the community and for the sake of all. This is not rational calculation, but something that "must happen of itself; it must be present in one's nature," and therefore it is something that is part of feeling and emotion, not reason. Dostoyevsky is appealing to the Russian idea of the commune, to a liberty that is combined with community, to an end to rational self-interest and calculation. He went to Europe and found what he expected to find—nothing like the admired ideal of the Russian aristocracy and some of the intelligentsia.

The essay—argued out of feeling and not at all systematically—was followed by *Notes from Underground*, a deep and insightful short novel which has in it, among many topics, one of the most profound and eloquent critiques of Western rationalism and modernity made in the mid-nineteenth century.

The novel has a short introduction in which Dostoyevsky tells the reader that, while this is fiction, the main character is someone who is part of Russian society now. The person, who has no name and is known in the canon as Underground Man, is representative. However, he is not Baudelaire's *flâneur* in Russian guise. Rather, he is a clerk, one of the anonymous many who compose society. He is full of contradiction and paradox. He is 40, angry at life, has consciousness to the point of having it limit his capacity to act, admits to lying occasionally, and is an unreliable narrator. He tells the reader he gets pleasure from his own degradation. Dostoyevsky manages to make the reader repelled by the man and fascinated by him at the same time. If Baudelaire's bourgeois is a "hero of modern life," UM is an anti-hero.

The critique of the West and its modernity comes in the first reflective part of the work. First, he challenges the notion that we are rational beings functioning according to reason and utilitarianism. The idea that humans act in their best interests is absurd for UM:

> What about all those millions of incidents testifying to the fact that men have *knowingly*, that is in full understanding of their own interests, put them in the background and taken a perilous and uncertain course not because anybody or anything drove them to it, but simply and solely because they did not choose to follow the appointed road, as it were, but wilfully and obstinately preferred to pursue a perverse and difficult path, almost lost in the darkness. This shows that obstinacy and self-will meant more to them than any kind of advantage.

We are far more complex than the West tells us. We choose for reasons which are not simply rational. Will and desire are often far more meaningful than rational calculation, something acknowledged by both economists and psychologists from the twentieth century. As UM tells us, there is something more important than his best interests to "almost every man—some best good...which is more important and higher than any other good, and for the sake of which man is prepared to go against all laws, against, that is, reason, honour, peace and quiet, prosperity—in short against all those fine and advantageous things—only to attain that primary, best good which is dearer to him than all else?" Rational man is a myth, one that ignores the depths and complexities of human nature.

The critique then turns to the fondness in the West in the mid-nineteenth century for what UM calls "systems," the establishment of laws of nature or society. He first attacks the idea in the works of Henry Thomas Buckle (1821–1862), a historian and philosopher very popular in Britain and Russia at the time, that there is a law of progress demonstrated throughout history. Buckle, he says, tells us that civilization renders humanity more peaceful, less violent, and war-like. That is nonsense. "Look around you," he says, "everywhere blood flows in torrents, and what's more, as merrily as if it were champagne." He cites both Napoleons, the United States Civil War, and the Danish War between Denmark against Prussia and Austria in 1864 over Schleswig-Holstein. We are so perverse that we ignore facts in order to support systems.

UM discusses the importance of suffering in several places. In the modern world we will eliminate suffering, he notes. But suffering and other feelings are part of what it means to be human: "Suffering—after all, that is the sole cause of consciousness." Dostoyevsky's argument with the West, part of the Slavophile critique, is that the world of rational man ignores the value of sentiment and suffering, the importance of desire and will, the primacy of feeling and brotherhood. UM tells us that "reason is only reason and satisfies only man's intellectual faculties, while volition is a manifestation of the whole of life, I mean the whole of human life *including* both reason and speculation."

Two other of his insights have become part of our reflections about human nature, society, and knowledge. In discussing what is happening in the West, the building and developments of cities and industry, UM suggests that we build, but not to achieve some edifice or road or finished project:

> Man loves construction and the laying out of roads, that is indisputable. But how is it that he is so passionately disposed to destruction and chaos? Tell me that? But on this subject I should like to put in two words of my own. Doesn't this passionate love for destruction and chaos (and nobody can deny that he is sometimes devoted to them; that is a fact), arise from his instinctive fear of attaining his goal and completing the building he is erecting? For all you know, perhaps it is only from a distance that he likes the building, and from close to he doesn't like it at all; perhaps he only likes building it, not living in it...

He likens us to ants that begin building a hill. And then he notes that humans are "fickle and disreputable." Perhaps we love the process more than the goal. Dostoyevsky's works, especially *Notes*, are valued highly by later existentialists, who saw in his musings much of what they were trying to say. When UM tells us we like the process and are afraid of the goal we are reminded of Camus' essay on the myth of Sisyphus and why he claims that Sisyphus can be happy.

And then there is an image which is unique at the time, though it will be fully understood two generations later, in the epistemological world of Einstein. In reflecting on the love of systems, UM uses the mathematical metaphor of $2 \times 2 = 4$ to refer to the closed self-referential system of current mathematics:

> In short, mankind is comically constructed... But $2 \times 2 = 4$ is nevertheless an intolerable thing. Twice two is four is, in my opinion, nothing but impudence. 'Two and two make four' is like a cocky young devil standing across your path with arms akimbo and a defiant air. I agree

that two and two make four is an excellent thing; but to give everything its due, two and two make five is also a very fine thing.

To adopt a system and follow it through life is to limit one. "After twice two is achieved there will of course be nothing left to do, much less to learn." Systems close off the realm of consciousness and creativity. They make you secure, but they also are confining.

Two generations later, the epistemological breakthrough of relativity tells us that there are other forms of mathematics and that we have frames of reference which are useful but they are not ontological realities.

UM tells his reader that believing in the Crystal Palace, that icon of modernity, is foolish. It is something that causes UM to stick out his tongue: "Destroy my desire, blot out my ideals, show me something better, and I will follow you." In the meanwhile, UM (and Dostoyevsky) find the modernity of the West unappealing, even inhuman in a number of ways.

There were other anti-moderns, including a number of groups, following Fourier, Owen, and others, who attempted to found utopian communities. Others found the urban life stressful and unappealing, and stayed in the countryside. Many individuals made their own separate peace and followed a lifestyle different from that of the industrial and political world of the time. And on the margins of the West, in Russia, the Balkans, and some places in southern Europe, modernity was seen as soul-destroying. These anti-moderns attacked what Hobsbawm called "the loneliness of liberalism" and emphasized community and tradition.

Modernity still is relentless in its development. Yet, the critique at the time mattered and many of the concerns of the religious, the anarchists, the craftspeople, and Dostoyevsky still resonate.

Chapter IX
Charles Darwin: The Mystery of Mysteries

1859 was a busy time for Charles Darwin (1809–1882). For some years he had been engaged in scientific research which led him to the conclusions that species were related to one another and that they changed over the course of time. He had hesitated to write about his findings, trying to accumulate as much material as possible to defend them. As well, he knew that his theories countered much that most people believed to be true in his Victorian England.

However, three years earlier, in 1856, several friends had encouraged him to publish, for the idea was being discussed in the scientific circles to which Darwin belonged as a member of the Royal Society. He was especially pushed to do so by his close friend, Charles Lyell, the most eminent geologist of the day. Lyell pointed to various publications, including one by a little-known naturalist, Alfred Russel Wallace (1822–1913), which suggested that there were relationships between the many varieties of species. Darwin then began writing material that he expected would result in a large set of books:

> Early in 1856 Lyell advised me to write out my views pretty fully, and I began at once to do so on a scale three or four times as extensive as that which was afterwards followed in my 'Origin of Species;' yet it was only an abstract of the materials which I had collected, and I got through about half the work on this scale.[135]

But science didn't stand still to wait for Darwin to finish. As Darwin states, "[M]y plans were overthrown, for early in the summer of 1858 Mr. Wallace, who was then in the Malay archipelago, sent me an essay [which]...contained the same theory as mine."

Darwin consulted Lyell and Joseph Hooker, a great botanist and also a close friend. There was the question of glory, of getting credit. And there was the associated matter of doing the correct thing as a gentleman and a scientist, which was much on the mind of the decent and troubled Darwin. His friends, noted scientists themselves, recommended that at the next meeting of the Linnean Society, the most prestigious group in the field of natural history, a combined contribution of material on the matter by both Wallace and Darwin should be presented. Darwin agreed. On July 1, 1858, three submissions were read. They were the essay sent to Darwin by Wallace, an 1844 paper by Darwin which was presented as "Extract from an MS. work on Species....sketched in 1839

---

135 Charles Darwin, *Autobiography* (Project Gutenberg, 1999), 27.

and copied in 1844," and a letter describing his ideas sent in 1857 to Asa Gray, Darwin's friend and a professor at Harvard University. The three documents were soon published in a journal of the Linnean Society. Surprisingly, given what shortly occurred, as Darwin said, "our joint productions excited very little attention."

What they did accomplish was to focus Darwin. After taking a holiday with his family, he decided to publish a shorter treatise than originally planned. It took 13 months to do so. The result was the one-volume *The Origin of Species by Means of Natural Selection or The Preservation of Favoured Races in the Struggle for Life*, published officially on November 24, 1859, a book that is easily one of the most influential and important scientific works ever produced in the West, on the same shelf with works by Ptolemy, Vesalius, Copernicus, Galileo, Newton, and, later, Mendel and Einstein.

Darwin would refer to his work as an "abstract," for he truly wanted to be scientifically comprehensive and longed to do the larger manuscript he had planned. Abstracts are usually several paragraphs in length. Darwin's "abstract" was 502 pages, which lets us know what he would have liked to present. However, the shorter work, tightly argued, written in a style which could be understood by an educated lay person, with no mathematics and only one diagram, was accessible to audiences far beyond those engaged in scientific research. As well, Darwin's writing in the book is engaging, very friendly to his readers, and, in places, elegant. It was widely read.[136]

Darwin came from a notable background in the world of English intellectual life. His two grandfathers were Erasmus Darwin (1731–1802) and Josiah Wedgewood (1730–1795). The former was a noted physician, poet, and inventor. The latter was a potter who was the first to turn pottery into a manufacturing industry. The two families intermingled. Darwin married his cousin Emma Wedgwood in 1839. They had many children and were a devoted and happy couple.

The family had money and the Darwins purchased a property in Kent, 18 miles from London, in 1842. They lived in Down House the remainder of their lives. Darwin was a gentleman scholar and scientist for the whole of his adult years. He had a variety of illnesses and as a result he and his wife lived quiet lives. By temperament as well, Darwin chose to remain on the sidelines in the controversies surrounding his work.

Prior to the publication of *The Origin of Species*, when the name Darwin was mentioned in literary circles it referred to his grandfather, Erasmus Darwin

---

[136] Some commentators have likened Darwin's style to some Victorian novelists, including Dickens and George Eliot.

(1731–1802), who proposed a number of theories about the origin of life, including speculation on the idea of spontaneous generation, as well as putting forward a "filament theory" in his *Zoonomia:*

> Would it be too bold to imagine that, in the great length of time since the earth began to exist, perhaps millions of ages before the commencement of the history of mankind...that all warm-blooded animals have arisen from one living filament, which the great First Cause endued with animality, with the power of acquiring new parts, attended with new propensities, directed by irritations, sensations, volitions and associations, and thus possessing the faculty of continuing to improve by its own inherent activity, and of delivering down these improvements by generation to its posterity, world without end![137]

Erasmus Darwin not only made contributions to science and literature, he became something of an inspiration for one of the great creations of literature about science in the nineteenth century, Mary Shelley's *Frankenstein* (1818), a work of speculative fiction and considered the first work in the new genre of science fiction.

Shelley tells us in the first sentence of her preface to the first edition, "The event on which this fiction is founded has been supposed by Dr. Darwin, and some of the physiological writers of Germany, as not of impossible occurrence." In the introduction to the better known (in the mid-nineteenth century) second edition of 1831, Shelley gives an account of the origin of the work, one well known to most students of modern literature. She notes that in listening to the conversations between Byron and Percy Shelley, the two men "talked of the experiments of Dr. Darwin," giving her the idea that sparked the work.

Shelley anticipates in her novel the experience of Charles Darwin on his voyage on the HMS *Beagle* from 1831 to 1836, his first foray into studying what would become evolution, a journey that made him as a scientist. She has Robert Walton, a ship captain, explain at the beginning of the novel that he is on "a voyage of discovery," something that was part of the excitement of the times. Not only that, but Walton, in going on his voyage, in his case to the North Pole, is motivated by "glory": "I preferred glory to every enticement..." Of course, Victor Frankenstein's obsession in creating life via what Shelley called "galvanism" mirrors Walton's quest.

In the 1830s, Charles Darwin knew he was involved in intellectual territory which would be enormously controversial. He, too, was on a voyage of discovery with extraordinary consequences.

---

[137] http://www.actionbioscience.org/evolution/buckeridge.html.

He needed three main ideas to support his research and develop a preliminary theory. First, he needed a new concept of time. If, as he began to conclude, species were mutable, challenging the prevailing belief in immutability, he needed to deal with the concept of time in a geological manner rather than in one based on the understanding of the Bible.

Most people in the West in the early nineteenth century still believed that the world was young. The famous and well-known calculation was that of Bishop James Ussher of the Church of Ireland. In 1650 he calculated that the day of creation was October 22, 4004 BCE, based on a literal understanding of the Old Testament. Others did their own calculations using the same method. They might have disagreed slightly with Ussher, but they still proclaimed a similar chronology.

The new study of geology fundamentally challenged this chronology (at least for some). James Hutton presented a number of new ideas to the Royal Society of Edinburgh in 1785, which became his study *Theory of the Earth*, published in 1788. One of these new notions was that the earth was very old, relative to what theologians thought. Thus began what has come to be called "deep time."[138] Then, it was thought of as long but not easily calculable. Today, geologists tell us that the earth began roughly 4.5 billion years ago.

Deep time meant that Darwin could explain the changes occurring in and between species as something that occurs very, very slowly over the course of a very long time. If species were mutable, the transformations were not at all sudden.

The second idea came out of Hutton's work and was fully presented to Darwin by Charles Lyell in the first volume of his *Principles of Geology* (1830).[139] It gave Darwin a model of change which helped him to theorize about what he observed on his voyage.

The prevailing geological theory about how the earth was formed was a version of the Creation story as told in the Bible. The myth was understood by geologists to be a metaphor—the earth was shaped in stages, perhaps six, perhaps a different number. This was called "catastrophism," and its most notable champion was the French geologist George Cuvier. He opined that there was a set of successive major events that explained geological evidence. He did not refer to theology, but he did believe in the fixity of species.

---

[138] A concept discussed by John McPhee in his *Basin and Range* (New York: Farrar, Strauss and Giroux, 1981).

[139] Darwin had that volume with him when he began his voyage on the HMS *Beagle* and read it then.

Hutton, Lyell, and other British scientists proposed a different idea, called "uniformitarianism." Changes did indeed occur. However, they did so very slowly as a result of nature working uniformly over very long periods of time. Moreover, these changes were not progressive in the valuative sense of that term. They occurred without considering either God or some outside metaphysical concept. And the laws which operated in the past are those still operating today. Time was immeasurable, the beginning was then unknowable, the end conformed to no special teleology. Our recorded history as human beings was but a tiny part of the life of the Earth. If we believed in the fixity of species because we could not observe the transformation, it was because our own part of Earth's history was but a tiny fragment of its life.

The result for Darwin was to provide him with a model for change which would guide his thoughts on the observations he made on his voyage, the voyage that led him to evolution. Janet Browne has stated, "Without Lyell, it could be said, there might not have been any Darwin..."[140] Darwin did not know Lyell when he read his *Principles of Geology*. The geologist would become a close friend and supporter after they met in 1836. Darwin always acknowledged his intellectual debt:

> The science of geology is enormously indebted to Lyell—more so, as I believe, than to any other man who ever lived. When [I was] starting on the voyage of the "Beagle", the sagacious Henslow, who, like all other geologists, believed at that time in successive cataclysms, advised me to get and study the first volume of the 'Principles,' which had then just been published, but on no account to accept the views therein advocated. How differently would anyone now speak of the 'Principles'! I am proud to remember that the first place, namely, St. Jago, in the Cape de Verde archipelago, in which I geologised, convinced me of the infinite superiority of Lyell's views over those advocated in any other work known to me.

The third main idea came to him suddenly, two years after the end of the *Beagle* voyage. In September 1838, while contemplating how to explain the mutability of species, he read the 1798 work of the gloomy Reverend Thomas Malthus, *An Essay on the Principle of Population*. Malthus put forth a simple mathematical proposition. He was witnessing the demographic growth that occurred in the early industrial revolution and sought an explanation. Humans increased as part of their ordinary lives. He suggested that humans increased geometrically. Food production, however, could never manage to match this increase, for it increased arithmetically. Hence, built into this ratio was human misery. The population was kept in check by famine, disease, war, and starvation.

---

[140] Janet Browne, *Darwin's Origin of Species: A Biography* (Vancouver: Douglas and McIntyre, 2006), 33.

Malthus desired to show that charity should be discouraged, for it only meant that those who were poor, weak, and destitute would breed even more and thus contribute further to misery. He attacked the existing poor laws as worse than useless. Indeed, Malthus even suggested that the misery that fell on the weakest in society was simply God's will.

Malthus was wrong in his assertion, but that didn't seem to matter at the time to those who opposed social reform or even to Darwin. It was a clearly powerful argument to those experiencing some of the devastating effects of the early industrial revolution in the country leading that development.

Darwin remarked on reading Malthus,

> In October 1838 [it was on September 28, 1838 that he recorded reading Malthus in a notebook]... I happened to read for amusement 'Malthus on Population,' and being well prepared to appreciate the struggle for existence which everywhere goes on from long-continued observation of the habits of animals and plants, it at once struck me that under these circumstances favourable variations would tend to be preserved, and unfavourable ones to be destroyed. Here then I had at last a theory by which to work.

The theory became natural selection. It was one which was in conformity with some of the other works of his time. Like Marx, Darwin emphasized struggle as a major part of life and transformation. Life was a struggle among species and also within species. Nature selected those most favourable to survive. However, nature was not reified. Nor was the process the result of the will or working of a deity. The process was part of a law of evolution and development. Like Lyell's uniformitarianism, the process had been going on for millions of years and was still going on.

There was one other important conclusion which Darwin would use, the idea of scarcity. Malthus' theory, if true, meant there would always be a scarcity of resources, and there would thus be a competition for those resources between and among species. Hence, in nature at every moment there was a kind of war among various creatures for the resources necessary to sustain and continue life. Tennyson is often quoted when describing Darwin's nature, from his *In Memoriam* (1850), Canto 56.

> So careful of the type?" but no.
>   From scarped cliff and quarried stone
>   She cries, "A thousand types are gone:
> I care for nothing, all shall go.
>
> "Thou makest thine appeal to me:
>   I bring to life, I bring to death:

> The spirit does but mean the breath:
> I know no more." And he, shall he,
>
> Man, her last work, who seem'd so fair,
>   Such splendid purpose in his eyes,
>   Who roll'd the psalm to wintry skies,
> Who built him fanes of fruitless prayer,
>
> Who trusted God was love indeed
>   And love Creation's final law—
>   Tho' Nature, red in tooth and claw
> With ravine, shriek'd against his creed—
>
> Who loved, who suffer'd countless ills,
>   Who battled for the True, the Just,
>   Be blown about the desert dust,
> Or seal'd within the iron hills?
>
> No more? A monster then, a dream,
>   A discord. Dragons of the prime,
>   That tare each other in their slime,
> Were mellow music match'd with him.
>
> O life as futile, then, as frail!
>   O for thy voice to soothe and bless!
>   What hope of answer, or redress?
> Behind the veil, behind the veil.

Darwin had then the basis of a monumental work. He hesitated, as we know, perhaps troubled by the implications of his ideas for his times, perhaps not wanting to enter into controversy, perhaps respectful of his beloved wife's religious beliefs.

Emma Darwin was aware of Charles' religious scepticism, which he wisely revealed to her before their marriage. It didn't matter to their relationship. She found him "transparent." She was also generally aware of his new ideas. Darwin did not think of himself as an atheist. When his friend Thomas Henry Huxley invented the term agnosticism in 1869, he called himself an agnostic. Interestingly, Emma called Charles a "materialist" in explaining his lack of religious belief. If materialism means a view rooted in data and in the realities of our life in whatever society we happen to be part of, that is so. Both Marx and Engels admired Darwin and his ideas.

The *Origin* opened with a brief historical sketch and a short introduction. Both were introducing Darwin and his grand idea. "Until recently," Darwin tells his readers, "the great majority of naturalists believed that species were immutable productions, and had been separately created." Moreover, he gives his adversaries some credit. "This view has been ably maintained by many authors."

Darwin goes on for about 10 pages tracing the concept from Buffon to his own day.

The introduction, less than three pages, prepares the reader for what is to come. It is written in a familiar tone, Darwin always being careful to make his ideas clear and using as much ordinary language as possible. This is no scientist talking to five other scientists, or warning his readers that they likely will not understand him. This readability is important, because the reader comes to think along with Darwin as the arguments developed and the book progresses.

He begins using the first person and his own experience:

> When on board the H.M.S. *Beagle*, as naturalist, I was much struck with certain facts in the distribution of the inhabitants of South America, and in the geological relations of the present to the past inhabitants of that continent. These facts seemed to me to throw some light on the origin of species—that mystery of mysteries, as it has been called by one of our greatest philosophers.

The philosopher (what we now call science was then often called natural philosophy in Britain) was John Herschel, the most respected scientist of his day, who made contributions to, among other fields, astronomy, botany, mathematics, and photography. In February 1836, Herschel wrote to Lyell, thanking him for sending Herschel a new edition of *Principles of Geology*. Herschel complimented Lyell for being bold enough to put forth his ideas on such a difficult subject, "that mystery of mysteries, the replacement of extinct species by others." Herschel, like many, didn't abandon God. Perhaps, he wrote:

> The creator...operates through a series of intermediate causes, and that in consequence the origination of fresh species, could it ever come under our cognizance, would be found to be a natural in contradistinction to a miraculous process.[141]

That part of the long letter was made public. For the well-read and discerning reader, Darwin is putting himself alongside the views of the two most respected scientists of his time.

Darwin lets the reader know that this publication is incomplete. But, he notes, it would take two or three more years to fully defend his thesis with the necessary material.

He emphasizes the importance of an understanding of what he calls modification and co-adaptation, which naturalists have not yet adequately explained.

---

141 http://www.blc.arizona.edu/courses/schaffer/249/Before%20Darwin%20-%20New/Cannon%20-%20Two%20letters.pdf.

Then Darwin prepares the reader for what is to come—developing the theory in the first several chapters. Especially important will be the third and the fourth, on "the Struggle for Existence" and "Natural Selection" (his caps).

In first mentioning the struggle for existence, Darwin adds that it is a struggle among all beings, following "the doctrine of Malthus applied to the whole of the animal and vegetable kingdoms." More of any species are born than can survive, hence any small variation might give some a better chance to survive the struggle, "and thus be *naturally selected*." Natural selection will result in the extinction of some forms, and results in "what I have called Divergence of Character" (his caps). The discussion continues in other chapters and concludes:

> Although much remains obscure, and will long remain obscure, I can entertain no doubt, after the most deliberate study and dispassionate judgment of which I am capable, that the view which most naturalists until recently entertained, and which I formerly entertained—namely, that each species has been independently created—is erroneous. I am fully convinced that species are not immutable; but that those belonging to what are called the same genera are lineal descendants of some other and generally extinct species, in the same manner as the acknowledged varieties of any one species are the descendants of that species. Furthermore, I am convinced that natural selection has been the most important, but not the exclusive, means of modification.

Darwin wisely begins the argument with "Variation under Domestication." It is a subject that can be understood readily by a lay reader. As well, it is one which most readers would have either experienced and/or already read about in numerous ways. After all, who in Britain did not know that horses are bred for strength and speed, dogs for special purposes, even plants for their yield and other qualities?

One of the most important matters regarding variation, Darwin admits, is not yet understood: "The laws governing inheritance are quite unknown." The work of Gregor Mendel will be done in the next decade. His most important paper, "Experiments in Plant Hybridization," was read in 1865 in Moravia and published the next year. It was the breakthrough essay on genetics, but it was virtually unknown, including by Darwin, until 1900, when several scientists replicated his work. It would not be integrated with evolutionary theory until 1930.[142] Out of this integration, the field of Evolutionary Genetics would develop.

Nonetheless, adaptation does occur in domesticated races, remarks Darwin, not because it is advantageous in any way to the plant or bird or animal, but because it is useful to the breeder: "The key is man's power of accumulative selec-

---

[142] See Ronald Fisher, *The Genetical Theory of Natural Selection* (New York: Dover Publications, 1958, originally published in 1930).

tion: nature gives successive variations; man adds them up in certain directions useful to him." Indeed, he gives the reader accounts of breeding for domestication in ancient times, citing, among others, passages in Genesis related to the colour of domestic animals, Roman times, early English history, and an old Chinese encyclopaedia. Certainly, the fact that domestic species are cross-bred results in the formation of new sub-species.

Still, Darwin reminds his reader (as he does many times in the book in various ways) of the main thrust of his ideas, "over all these causes of Change I am convinced that the accumulative action of Selection... is by far the predominant Power" (his caps).

Now Darwin will get to the important subject of "Variation under Nature," on which his thesis depends. He notes he will be looking at small differences which, over the course of time, will matter greatly. It is a short chapter which lets the reader anticipate what will be elaborated upon later in greater detail. The most important concept he introduces is that of Divergence of Character (his caps). He calls it a principle which will explain "how the lesser differences between varieties will tend to increase into the greater differences between species." Hence, varieties matter. Sometimes, it is very difficult to distinguish a variety from a species. However, varieties, he notes, at some point "tend to become converted into new and distinct species." Moreover, in nature forms of life which become dominant "tend to become still more dominant by leaving many modified and dominant descendants." He is very careful here in his language. Things "tend" to happen. There are no "musts" or "necessities" or "inevitabilities." He will offer no numbers or formulae in the story he is telling about how evolution operates.

The next two chapters, "Struggle for Existence" and "Natural Selection," are keys to the argument. In them Darwin will discuss many matters central to evolution, getting his readers to understand his main ideas and hopefully to accept that this is a possible explanation of the bases of species.

He starts on a theoretical level. How do varieties, which he terms "incipient species," turn into species, which are able to survive and are distinct? How do groups of species, called genera, come into being? All this arises out of "the struggle for life." Any useful variation will help survival and will be transmitted by inheritance: "I have called this principle, by which each slight variation, if useful, is preserved, by the term of Natural Selection." Indeed, praising the workings of nature, Darwin poetically asserts that "Natural Selection...is immeasurably superior to man's feeble efforts, as the works of Nature are to those of Art."

The chapter is filled with images of destruction, struggle, extinction, war, and competition. He notes that as we delight in the beauty of songbirds, those birds feed on insects and seed, and are themselves attacked by birds of prey.

Look at the songbird and nature appears benign. Go beyond appearance and nature is in reality rough and harsh. The destruction of some species is necessary, because our world cannot contain all that are developed. He gives the famous example of the elephant, the slowest breeder of all. Some must die in competition, for if none died and breeding occurred more rapidly, there would be 15,000,000 elephants on our earth five centuries hence.

A single fly has hundreds of eggs, a plant might have many seeds. Though there will be much destruction, the producer of many has an advantage in that some will survive: "Lighten any check, mitigate the destruction ever so little, and the number of species will almost instantaneously increase...."

Climate is significant. If the food supply is reduced as a result of climate changes, it increases the struggle for existence and heightens the competition between individuals and species. Thus, in the slow processes of natural selection, it is far better to belong to a species whose numbers are larger than those of its enemies, to assist survival. It is worth considering whether this idea was something understood easily in Victorian times. It was common to have children die in infancy, to have some survive and others not. Darwin and Marx, both of whom had large families, knew first-hand, as Marx lost four of his children in infancy or childhood, and Darwin three.

He suggests that it is superficial to attribute the population of plants in any bank of what he calls a "country," meaning an environment, to chance. Over the course of centuries, environment, too, has changed according to the principles of natural selection. There will be a "struggle" that is "severe," he notes more than once, the "competition" will also be "severe." It is a "great battle of life." He insists on the Malthusian geometrical/arithmetical reality: "[E]ach at some period of its life, during some seasons of the year, during each generation or at intervals, has to struggle for life, and to suffer great destruction."

The language used in this chapter and throughout the work is as powerful in its images as some of those used by Marx and Engels. The idea of conflict is no less significant in Darwin's nature than it is in Marx's class struggles in the dialectic. The process for both is part of how humans develop and survive and triumph.

At the end of the chapter, Darwin tries to offer his readers some consolation. The struggle persists, he writes, but "we may console ourselves with the full belief that the war of nature is not incessant, that no fear is felt, that death is prompt, and that the vigorous, the healthy, and the happy survive and multiply." This sentence goes against much of what Darwin said in the rest of his reflections on "Struggle for Existence." It is recorded that Darwin enjoyed reading Wordsworth at one time in his life. But his nature is not that of Wordsworth. It is Tennyson's, red in tooth and claw.

He opens the chapter on "Natural Selection" with a discussion on variation—and a definition: "This preservation of favourable variations and rejection of injurious variations, I call Natural Selection." Modifications matter. Every small modification which acts to favor individuals of any species because it assists in adapting to conditions would "tend to be preserved." Hence, natural selection acts in "the work of improvement." The language Darwin uses is that of the age of progress, for example "improvement" or "profitable variations," but it should be understood that the progress is not valuative, but normative.

Indeed, in providing the reader with examples of how natural selection works, Darwin says, "I must beg permission to give one or two imaginary illustrations." And he then does so, taking one example from the animal world and another from plants. Darwin was subsequently challenged by commentators who suggested that his logic was in error if he was doing science via imagination.

Darwin was defended by John Stuart Mill, though the defence also weakened any claim that natural selection had been conclusively proven. In editions of his *System of Logic* (first published in 1843) completed after Darwin's work came out, Mill included a footnote on "Mr. Darwin's remarkable speculation" in the context of Mill's discussion of the nature of hypothesis. Mill thought hypotheses important as part of the way discoveries were made:

> Mr. Darwin's remarkable speculation on the Origin of Species is another unimpeachable example of a legitimate hypothesis. What he terms "natural selection" is not only a *vera causa*, but one proved to be capable of producing effects of the same kind with those which the hypothesis ascribes to it; the question of possibility is entirely one of degree. It is unreasonable to accuse Mr. Darwin (as has been done) of violating the rules of Induction. The rules of Induction are concerned with the conditions of Proof. Mr. Darwin has never pretended that his doctrine was proved. He was not bound by the rules of Induction, but by those of Hypothesis. And these last have seldom been more completely fulfilled. He has opened a path of inquiry full of promise, the results of which none can foresee. And is it not a wonderful feat of scientific knowledge and ingenuity to have rendered so bold a suggestion, which the first impulse of every one was to reject at once, admissible and discussible, even as a conjecture?[143]

Given that Darwin called his work an "abstract" and wished he could have had more time and space to prove himself, Mill's contention helps. Many would work very hard to either prove or disprove the *Origin*.

Darwin goes on to discuss the important concept of divergence, which he noted is of great importance to his theory. Diversification matters in terms of survival and development. The greater the diversification, the greater the advantag-

---

143 https://www.gutenberg.org/files/27942/27942-h/27942-h.html#toc49, footnote 164.

es in the conflicts in nature. The principle is that "the greatest amount of life can be supported by great diversification of structure." It gives advantages to life forms in the same region. He believes that "a set of animals, with their organisation but little diversified, could hardly compete with a set more perfectly diversified in structure." Diversification and modification are related.

Darwin at the end of this discussion sums up the chapter. By this moment in the book the main ideas have been stated, often more than once in different contexts, so as to both help the reader and give more support to the argument. He here attacks openly the "view that each species has been independently created. I can see no explanation of this great fact in the classification of all organic beings." Rather, the explanation is best done via inheritance and natural selection, knowing that divergence is central.

He then, in the last (long) paragraph in this central chapter, offers a simile which borrows from ancient biblical thought. All the beings of the same class, he suggests, can be likened to "a great tree." The twigs represent what exist. The tree grows. The twigs branch out and some twigs die. Only two or three of the twigs which existed when the tree was but a bush grow into branches and they bear other branches. So it is with ancient species. Only a few have descendants which are modified and diverse. We only know those who died by examining them as fossils.

There is a "great Tree of Life," which fills with its dead and broken branches the crust of the earth, and covers the surface with its ever branching and beautiful ramification. The tree of life[144] was a powerful image, as was, of course, the Tree of Knowledge. Now, in explaining the one diagram in the book which is in this chapter, Darwin is associating the trees of life and knowledge also with time.[145]

It could be argued that most of the nineteenth century was historicist. Time mattered. Understanding change through time was the key to laws and knowledge. After all, nationalists needed a past to justify the present; socialists used the development of history to understand the present and to try to change it; geologists changed how we thought about time. Now, biology and human nature, via Darwin's hypotheses, came to be understood via slow change through a very, very long time.

In the concern about Becoming and transformation in the nineteenth century there was also a desire to order time as well to understand its laws. Solar time changed into standard time as the century progressed. The railway, so important

---

144 Eitz Chaim in Hebrew.
145 Browne, op. cit., 74.

to Cerdà and others in introducing modernity, became the impetus for fixing the clock, and was especially useful to the new industrial factories and trade. The first use of standard time was in Britain, by the railways, in 1847, using Greenwich Mean Time. It was known then as Railway Time.

The first suggestion that there be a worldwide organization of time zones was proposed by the Italian mathematician and politician Quirico Filopanti (the pseudonym of Giuseppi Barelli) in 1858. He suggested in his book *Miranda* that there be 24 time zones, with Rome at the center. This idea and variations of it were discussed at several international conferences that followed. In 1884 at the International Meridian Conference, a universal system beginning at Greenwich, England was adopted. The seventeenth century explored the heavens and their order and gave us laws. The nineteenth century explored change through time in an effort to also give us laws, as well as to find an order appropriate to modernity.

Darwin followed these first four central chapters with more evidence, reflections, and commentary. He elaborated on the laws of variation and spent a chapter each on instinct and hybridism. Two chapters were on geology, the record of how it occurred and its relationship to organic beings. Two others focused on geographical distribution. Another investigated morphology, embryology, and rudimentary organs.

Chapter 6, "Difficulties on Theory," was important as much for Darwin's acknowledgment of difficulties, objections, and problems as for its content. He was complimented by his critics for admitting the theory needed more underpinning. He also notes that the record is incomplete; more must be done in science. Indeed, he points out that some of the issues are "very grave." Still, he establishes the evidence tending in favor of modification, diversification, struggle, and natural selection. In the penultimate paragraph he states, "On the theory of natural selection we can clearly understand the full meaning of that old canon in natural history, '*Natura non facit saltum*'" (Nature does not make leaps).

In his conclusion, Darwin permits himself some wider speculation and reflection. He knows what he is doing, and he knows that there will be controversy and a possible reorientation of the whole of biological science. But he states, with some modesty:

> When the views entertained in this volume on the origin of species... are generally admitted, we can dimly foresee that there will be a considerable revolution in natural history....
> In the distant future I see open fields for far more important researches. Psychology will be based on a new foundation, that of the necessary acquirement of each mental power and capacity by gradation. Light will be thrown on the origin of man and his history.

He asserts that his views do not demean living beings, but "they seem to me to become ennobled." All living forms are interrelated and there can be a future in which "all corporeal and mental endowments will tend to progress towards perfection." The language is that of the idea of progress, though the content is really not. Perfection means better adapted, more diversified, better able to survive in a world with scarce resources.

The last paragraph is lyrical, for Darwin tells us in this work and in everything else he writes about the beauty and wonder of the nature he studied. It has been much quoted, deservedly:

> It is interesting to contemplate an entangled bank, clothed with many plants of many kinds, with birds singing on the bushes, with various insects flitting about, and with worms crawling through the damp earth, and to reflect that these elaborately constructed forms, so different from each other, and dependent upon each other in so complex a manner, have all been produced by laws acting around us. These laws, taken in the largest sense, being Growth with reproduction; Inheritance which is almost implied by reproduction; Variability from the indirect and direct action of the conditions of life, and from use and disuse; a Ratio of Increase so high as to lead to a Struggle for Life, and as a consequence to Natural Selection, entailing Divergence of Character and the Extinction of less improved forms. Thus, from the war of nature, from famine and death, the most exalted object which we are capable of conceiving, namely, the production of the higher animals, directly follows. There is grandeur in this view of life, with its several powers, having been originally breathed by the Creator into a few forms or into one; and that, whilst this planet has gone circling on according to the fixed law of gravity, from so simple a beginning endless forms most beautiful and most wonderful have been, and are being evolved.

It is interesting to also reflect upon what is not in this highly important and influential work. Darwin, with few exceptions, including in the last paragraph where he borrows from both biblical studies and poetry, does not mention a creator who creates species. He is clear throughout that "I view all beings not as special creations." It may be beautiful and it may be wondrous, but it is not divine. Nature has its ways and its laws.

What Darwin did not do (even though mentioning the law of gravity, associated in all educated minds with Newton) was what Newton did at the end of his great masterpiece, the *Philosophiæ Naturalis Principia Mathematica* (1687). Newton concluded with sincerity and conviction that "this great system of the sun, planets, and comets, could only proceed from the counsel and dominion of an intelligent and powerful Being.... The Being governs all things." Newton discovered God's laws. Darwin discovered those in Nature.

As well, Darwin does not opine on the first origin of life. Like Lyell in the geology of the time, that data seems too remote and lost to be considered. He is also mute on the origins of humanity. In this work he does not consider our beginnings, though that matter is one that will be on the minds of virtually all commentators, both supporters and critics. Later, in 1871, Darwin will publish *The Descent of Man, and Selection in Relation to Sex*, in which he stated, "The sole object of this work is to consider, firstly, whether man, like every other species, is descended from some pre-existing form; secondly, the manner of his development; and thirdly, the value of the differences between the so-called races of man."

Even accounting for his modesty, his admission that the theory needed more research, that it fell into the category of, as Mill later stated, hypothesis, and Darwin's reticence to go into the origins of human beings, it quickly became the most controversial work of its time.

The dissenters in science were headed by Adam Sedgwick and Richard Owen. Sedgwick had been Darwin's teacher of geology at Cambridge and the young Darwin had accompanied him on a field trip to Wales. They became friends. Darwin had a copy of the *Origin* sent to Sedgwick and on November 24 Sedgwick wrote to his former student, first thanking him for sending the book and then, "I have read your book with more pain than pleasure." He told Darwin that he had "*deserted*... the true method of deduction." He attacked Darwin's understanding of natural selection.

For Sedgwick, "I call causation the will of God: & I can prove that He acts for the good of His creatures." Nature, he believed, has moral purpose and metaphysical content, which Darwin denied. Moreover, he implies a teleology as well. He concludes that "I humbly accept God's revelation of himself both in His works & in His word." Sedgwick, for all his problems with the book, nonetheless wrote as a friendly critic and worked to maintain the personal tie.[146]

Darwin replied two days later, thanking Sedgwick for his frank criticism: "I grieve to have shocked a man whom I sincerely honour." Still, he noted that if he is in error "I shall soon be annihilated." He, too, was enormously respectful in the short letter, indicating that time and others will test the theory.

Richard Owen was Britain's leading anatomist and superintendent of the natural history collection for the British Museum. He was also known as someone with a fiery temperament and great ambition. At first, he complimented Darwin and they had some friendly discussions. Then, he wrote a scathing anonymous review in the April 1860 *Edinburgh Review*. And from this moment on Owen became a major opponent of Darwin's ideas, attacking the work from many sides, theological, factual, logical, and theoretical. His attacks made Darwin something he rarely became, upset. He wrote to Lyell,

---

[146] *The Correspondence of Charles Darwin*, vol. 7, 396–397.

> I have just read the Edinburgh (review of the *Origin*), which without doubt is by Owen. It is extremely malignant, clever & I fear will be very damaging. He is atrociously severe on Huxley's lecture, & very bitter against Hooker. It requires much study to appreciate all the bitter spite of the remarks against me....He misquotes some passages....
> It is painful to be hated in the intense degree with which Owen hates me.[147]

Many clerics, Anglican and others, challenged Darwin's conclusions, often from dogma. "We know all there is to know about it," said one. "God created plants and animals and man out of the ground."[148] Still, there were others who had already accepted Lyell's geology and the ideas of uniformitarianism. Hence, some like Charles Kingsley and Baden Powell took evolution and linked it with the idea of God's plan, one that had a principle which saw nature as evolving. The result was that, while evolution mainly was contested, there was something of a debate within the established church about it.

Protestant fundamentalists, then and now, simply reject both Lyell's geology and Darwin's biology. The Bible is to be understood literally and these ideas are patently false. This continues to the present, especially in some parts of the United States.

There was official silence from the Vatican, at a time when the Church was attacking every form of liberalism, socialism, democracy, and the sole reliance on reason to understand the world, scientific and moral. The result was that in individual countries and dioceses evolution might be challenged, but not officially by the Church itself. The *Origin* was never placed on The Index of Forbidden Books (ended in 1966), which did include some works by Balzac, Mill, George Sand, Zola, Bergson, and John William Draper's 1874 study, *History of the Conflict between Religion and Science*.

There were many defenders, led by Lyell, Joseph Hooker, and Thomas Henry Huxley, who came to be known as "Darwin's Bulldog." Darwin did not stay totally on the sidelines, but neither did he enter the fray in any active public manner. Largely, he stayed at home and wrote numerous letters which helped to organize the argument in favor and gave support to those on his side.

Huxley, a biologist with a specialty in anatomy as well as an anthropologist, a friend of Darwin, recalled his response on first reading the Origin:

> The 'Origin' provided us with the working hypothesis we sought. Moreover, it did the immense service of freeing us for ever from the dilemma – Refuse to accept the creation hypothesis, and what have you to propose that can be accepted by any cautious reasoned? In

---

147 Darwin, *Correspondence*, Vol. 8, 154.
148 Gertrude Himmelfarb, *Darwin and the Darwinian Revolution* (New York: Anchor Books, 1962), 301.

1857 I had no answer ready, and I do not think that anyone else had. A year later we reproached ourselves with dullness for being perplexed with such an inquiry. My reflection, when I first made myself master of the central idea of the "Origin" was, 'How extremely stupid not to have thought of that!'[149]

For Huxley (who, as mentioned above, in 1869 coined the word agnostic)—and many others—it was a "eureka" moment. Now, they had a frame of reference which could challenge the theological ideas. He and other defenders sometimes took issue with a part of the theory, or sometimes looked for data which would test some of Darwin's ideas. They did not swallow it whole. However, they did accept the main concepts and hypotheses.

Huxley was central to the most famous public debate on evolution, one which became part of the tale of its reception everywhere, an event that became part of the history of science.

On June 30, 1860 an exchange took place between Huxley and Bishop Samuel Wilberforce at a public meeting of the British Association for the Advancement of Science, at Oxford, Wilberforce's diocese. A crowd attended, estimated between 700–1000, so large that the venue had to be changed. The main lecturer was the American John William Draper, discussing Darwin's ideas related to the intellectual history of Europe. His contribution was boring and has been forgotten, but what happened afterwards electrified the scientific community and society in general.

Wilberforce, a charismatic speaker, rose and then attacked Darwin and Huxley on both anatomical and theological grounds. Most believe he received much of his argument from his friend Richard Owen, who was present. Then, at the end of his remarks, he turned to the audience and asked whether women also, like men, were thought to be derived from the animal kingdom. He then addressed Huxley. Was it from his grandmother's or his grandfather's side that Huxley was descended from apes?

Huxley realized the possibilities of an answer and turned to his neighbor, the surgeon Sir Benjamin Brodie, and whispered: "the Lord hath delivered him into my hands."

Huxley rose to reply and first defended the *Origin* as he had already done in several reviews, including one in the *Times*, the most influential newspaper of its time. He was careful to note that Darwin did not speak about the descent of man

---

[149] Thomas Henry Huxley, "On the Reception of the 'Origin of Species," in *The Life and Letters of Charles Darwin, Including an Autobiographical Chapter*, ed. Francis Darwin (London: John Murray, 1887), volume 2, 179–204 (quote is on v2, 197), https://profjoecain.net/how-extremely-stupid-thomas-henry-huxley/.

in his study, rather that humans and apes might be descended from a common ancestor.

At the close Huxley answered Wilberforce's question. He recalled it in the following manner:

> I asserted...that a man has no reason to be ashamed of having an ape for his grandfather. If there were an ancestor whom I should feel shame in recalling, it would rather be a *man*, a man of restless and versatile intellect, who, not content with an equivocal success in his own sphere of activity, plunges into scientific questions with which he had no real acquaintance, only to obscure them by an aimless rhetoric, and distract the attention of his hearers from the real point at issue by eloquent digressions and skilled appeals to religious prejudice.[150]

There is no reliable verbatim account of the exchange. Hooker, in a letter to Darwin about the event written two days later, stated: "Huxley answered admirably and turned the tables.... The battle waxed hot. Lady Brewster fainted, the excitement increased."[151] Huxley, about two months later, wrote to a friend about what he said:

> then, said I, the question is put to me would I rather have a miserable ape for a grandfather or a man highly endowed by nature and possessed of great means and influence and yet who employs those faculties for the mere purpose of introducing ridicule into scientific discussion—I unhesitatingly affirm my preference for the ape.[152]

The common understanding was a shortened version of Huxley's letter. Moreover, it produced a great stir in the audience, so much so that it was reported everywhere. It seemed to be the moment when the battle between science and religion was joined publicly. And Huxley was generally seen to have won. From this moment on, in Britain and then in Europe, there was something known as Darwinism.

Huxley continued to be Darwin's defender. In 1863 he published a work on the subject, *Evidence as to Man's Place in Nature*, which was the first work to directly address the descent of humans in Darwinian terms. He successfully attacked the anatomical work and conclusions of Richard Owen, who tried to prove the uniqueness of human beings. To Huxley, we are related to gibbons, orang-utans, chimpanzees, and gorillas.

---

150 Himmelfarb, op. cit., 291.
151 Darwin, *Correspondence*, vol. 8, 270.
152 Ibid., 271–72.

Darwin had many other important supporters, most notably Lyell and Joseph Hooker. Also in 1863, Lyell published *The Antiquity of Man*, a major work in both geology and anthropology, discussing the early stages of human development. Though Lyell was hesitant to go as far as Darwin in fitting humans into natural selection and believed that humans did have divine origin, he nonetheless gave him great scientific and personal assistance.

Hooker was a great botanist, the director of Kew Gardens near London, then the most important botanical research center in the world. His research included showing how Darwin's hypotheses worked in plants. He was Darwin's closest friend and the first major scientist to publicly support natural selection.

Karl Marx died in 1883, 11 months after the death of Darwin. In his eulogy at the gravesite in Highgate Cemetery, Engels, summing up Marx's contributions, mentioned only one other thinker: "Just as Darwin discovered the law of development of organic nature, so Marx discovered the law of development of human history."[153]

Engels read the *Origin* shortly after it was published and described it as "absolutely splendid." He praised the attempt to see evolution in nature. In 1861, Marx, in a letter to the socialist Lassalle, said, "Darwin's work is most important...in that it provides a basis in the natural sciences for the historical class struggle."[154]

Marx and Engels had some reservations about the analogy, for evolution was not revolution and species were not classes, but they had great respect for what Darwin was doing in an age where there was a search for the laws of development in society and nature. Marx sent an inscribed copy of the third edition of *Das Capital* to Darwin as a token of respect. Darwin replied politely, thanking Marx for his "deep and important [work] of political Economy." Still, Darwin did not fully read the work, perhaps because it was in German, perhaps because this was not a subject he ever studied—only the first 102 of the 822 pages were cut.[155]

There were differences, especially in the context of the times. Evolution, indeed, was not revolution. Darwin fully embraced uniformitarianism. Marx and Engels stressed the need for abrupt change, what the geologists styled as catastrophism. However, it could be argued that Marxism really had an evolutionary side. The Marxists identified only four previous societies—primitive communism, ancient, feudal, and the new bourgeois order, in spite of all that had previously been written about the history of humanity. Hence societies needed to evolve, as

---

[153] Tucker, op. cit., 681.
[154] *The Cambridge Companion to the 'Origin of Species'*, ed. Michael Ruse and Robert J. Richards (Cambridge University Press, 2009), 306–7.
[155] Ibid., 310.

Marx noted in his *Eighteenth Brumaire* and elsewhere, before they were ready for the revolution which transformed them. In geological terms, Marxism was uniformitarian until a moment when humanity was ready for a revolution, a catastrophe.

Marxists were teleological until the moment when there would be no contradictions and the laws of development would be transcended. Conflict would end. Not so for Darwin. There was no teleology, there was process, but it could not be described as progress in the valuative sense of that word. The struggle for existence and natural selection simply continued. And it would continue among species into the future.

Philosophical speculation had a purpose for Marx and Engels—to change the world morally and for the better. Darwin was interested in understanding nature. That would be enough. Of course, both Marx and Darwin did profoundly change our understanding of human nature and how societies, in political economy and in nature, function.

Sigmund Freud, who also changed how we think about ourselves, wrote in 1917 a reflection on how the narcissism of human beings, as he called it, was "severely wounded by the researches of science." Humans had for some time thought of the Earth as stationary, with the heavenly bodies circling it, a way of regarding ourselves as central to the universe. Then came Copernicus and others to destroy that "illusion." This was what Freud called the cosmological blow to our ego. Now, as some scientists tell us, we are on a third-rate planet revolving around a fourth-rate sun in a minor galaxy.

Still, humans continued to regard themselves as special creations, destined to dominate nature and the whole of the animal kingdom. Humans were even given immortality in some theologies, and they had a divine quality. The second blow to our narcissism came with Darwin, said Freud. Now we are part of the animal kingdom, and part of natural development. In the Victorian period this issue was discussed in such terms as whether we resembled apes or angels. Darwin, though he was very careful, nonetheless came out on the side of the apes (though he, and many others, thought we had the capacity to behave like angels if we chose to do so). In his 1871 work *The Descent of Man*, he concluded:

> Man may be excused for feeling some pride at having risen, though not through his own exertions, to the very summit of the organic scale; and the fact of his having thus risen, instead of having been aboriginally placed there, may give him hope for a still higher destiny in the distant future. But we are not here concerned with hopes or fears, only with the truth as far as our reason permits us to discover it; and I have given the evidence to the best of my ability. We must, however, acknowledge, as it seems to me, that man with all his noble qualities, with sympathy which feels for the most debased, with benevolence which extends not only to other men but to the humblest living creature, with his god-

like intellect which has penetrated into the movements and constitution of the solar system—with all these exalted powers—Man still bears in his bodily frame the indelible stamp of his lowly origin.

The *Origin* transformed the whole field of biology, as well as influencing other disciplines. From this moment forward it—along with genetics in the twentieth century—became the center of biological research.

Most importantly, Darwin's telling of our story gave impetus to countering the traditional theological version. His theory has become the most plausible way to understand humanity's development and nature without resorting to sacred texts, supernatural beings, or divine intervention. Though Darwin did not deal with them, he knew that his work would challenge tradition and established religion. It contributed mightily, for good or not, to the secularization of life in modernity. Further, it changed the way we think about human nature forever.

As it turns out, Darwin was buried in Westminster Abbey. By the time of his death, in 1882, he was seen as a great figure by most in the country, an Englishman who helped to conquer nature in his own way, and was revered nationally and internationally. By 1882, a great agnostic who lived a modest life could be honored in this way. Darwin, in the short *Autobiography* he wrote for his children, ended his recollections with characteristic understatement:

> Therefore, my success as a man of science, whatever this may have amounted to, has been determined, as far as I can judge, by complex and diversified mental qualities and conditions. Of these, the most important have been—the love of science—unbounded patience in long reflecting over any subject—industry in observing and collecting facts—and a fair share of invention as well as of common sense. With such moderate abilities as I possess, it is truly surprising that I should have influenced to a considerable extent the belief of scientific men in some important points.[156]

---

[156] *Autobiography*, op. cit., 37.

Epilogue – Reflections

Modernity is still with us. The issues opened up and considered in 1859 have not gone away. The 1859 poets, artists, scientists, philosophers, and historians can still inform us today.

1) How do we understand and respond to new modes of living resulting from the new political discourse, urban life, and a new economic reality?

The new is confronting us daily. Urban life is at the center of most everything that is occurring. Cities are so big and so important that they would hardly be recognized by someone writing in 1859. Our political discourse still makes Mill, Mazzini, Marx, and others extraordinarily relevant, as we grope with issues related to nationalism, human rights, and class. Our economic reality is complex, yet it deals profoundly with matters related to how wealth is distributed and the matter of state control.

What is totally different is that we have lived through the twentieth century and much of the downside of humanity. Yes, we have witnessed the growth and expansion of liberal democracy, the extension of the vote, and the acceptance of women and people of color as free citizens. But we have also experienced two world wars coming out of a Europe which mistakenly saw itself as in the forefront of civilization and progress, totalitarianism and the power of the modern states, genocide, and casual discrimination and racism.

Human rights have advanced, especially after the Second World War, but they are being attacked by modern states, openly in some EU states, easily in Russia and other authoritarian states, subversively in many so-called democracies which are tracking their citizens and fostering surveillance capitalism. It was a bone-fide liberal, José Ortega y Gasset, who in 1930 said that the greatest danger is the state, echoing John Stuart Mill, whose critique of state power now seems prophetic. It is a liberal economist, Thomas Piketty, leading the attack on the way we distribute our great wealth, because it effectively makes us less democratic. The anarchists were right in their concerns, if not in their solutions.

2) Is it necessary to write, paint, and do music differently in order to capture the modern experience and reality? Do we need a new language and philosophy to comprehend what is happening and what to do about it?

All the 1859 writers and artists answered these questions with a resounding "yes." They recognized that what they inherited was important, but it was not sufficient to deal with the world they were then experiencing. Modernity, as Baudelaire, Cerdà, and Marx insisted, was different. Hence, they had to find ways of comprehending it, which meant inventing new modes of analysis and new categories of understanding.

What is common among the 1859 thinkers is the need to transcend what was given to get to a sense of properly analyzing the world of Becoming. Marx and Engels contributed to doing social analysis differently, Darwin transformed biology and human nature forever, Baudelaire insisted on re-inventing poetry, and Manet struggled to find in his art a way of getting at the beauty and experience of modernity. Indeed, Burckhardt approached history totally differently from those in Berlin following Ranke, and his work is a prelude to much that we do in the discipline today.

We have witnessed other attempts to capture the modern experience, experiments, as it were, in epistemology and aesthetics. Science is continuing to invent ways of comprehending our world, in addition to the discovery of data. Artists have transformed our sense of how we see and understand—the work of Cubists or Surrealists, and many others, is as revolutionary as was relativity for physics. Most significantly, our view of human nature has changed and continues to do so. We are closer to Dostoyevsky than to Mill; we comprehend Baudelaire far easier than did his contemporaries. Darwin changed not merely how we came to be; he transformed how we think about ourselves.

3) How do we understand identity and aspirations now that the old theology and the central position of churches are challenged?

Theology plays a lesser role in Europe than it has since the fourth century. It exists, and churchmen still argue about it and provide one dogma or another, but it doesn't play an important role in the world of philosophical discourse or in the modern state. Churches in Europe are closing; many are empty.

Churches have a role, but it is mainly social. They are still used by many to mark and celebrate important events in the life-cycle—births, the coming to adulthood and responsibility, marriages, deaths. Church holidays are still kept, especially Christmas and Easter, though for many they are simply time off from work and for family events. Stores sometimes close on other days, Pentecost for example, but it is merely a holiday for most, not an observance.

The modern state has taken over, for good or ill. Nominally Christians, most Europeans view themselves as members of a nation and/or a state. And the modern state, as Hobsbawm and others have pointed out, has invented tradition, so that its founding and important celebrations are new holidays.

In some instances, states attempt to use the national church as part of their cloak of power, as in Russia, Turkey, and even the United Kingdom, where the monarch is the head of the Anglican Church. In others, the Scandinavian states for example, the church has little authority and much less presence that it would have had in 1859. France and some other states are relentlessly secular. Most significantly, churches have lost their power and control over education. There is indoctrination in the educational system of most states, but it is political dogma that is taught, not anything theological. Schooldays often open with some observance of loyalty to the state, be it a statement and/or a song. The flag has replaced the cross.

4) What is to be made of the new industrial society? What is just? What ought to happen with the new wealth and new classes of people?

Industrial society does what Dostoyevsky said might be the nature of modern development. We build it and then we destroy it so that we can build it anew in a different way. Today's industry would be unrecognizable in 1859. And yet the issues remain much the same.

The distribution of wealth in our industrial society is recognized as unfair, though many states do little about it. This may be because modern states are, as Marx and other socialists believed in 1859, part of the structure of power controlled by the wealthy. When the financial crisis of 2008 broke out, there was panic. Yet states bailed out the wealthy, few of those who were responsible for the crisis paid any price, and the banks were bailed out with the money of the middle class and the poor.

Europe is more equal than it was in 1859, but it is not even roughly equal in condition or in economic circumstance. There is wealth, but there is deep poverty even in wealthy states like France and the United Kingdom. There is resentment in parts of Europe towards the bureaucracy of the EU, but the EU institutions are far more representative of the wealthy class in Europe than the middle or the poor. Those who said in 1859 that economics determines politics may have been correct.

There is still much to learn from 1859 in today's world. Mill tells us about dignity, human rights, and the need for individuals to be subjects. Marx and Engels warn about the power of money and the lack of compassion of the wealthy class. Turgenev relates how eruption can occur which will transform lives and end the old regime. The street remains a place of contention in many places.

5) Has nationalism become the new way of organizing our political and civic culture? What are the consequences of the new national identity?

Socialism has made profound contributions to better the lives of Europeans, especially in the development of democratic socialism after the Second World War. People have unions which are effective, health programs which are national, education which is accessible, pensions to help them in old age, and much more.

Liberalism has also made important contributions. It is Liberalism which has defended and forwarded universal human rights and a fair justice system. Women, people of color, those with a sexual orientation different from heterosexuality, all have made gains in becoming more equal and having opportunity.

Conservatism has contributed a sensible attitude towards the process of change, a focus on society as far more important than politics, and a belief in the value of tradition and institutions.

But it is nationalism that still carries the day. It has not disappeared. Rather, with the decline in traditional religion combined with the growth and power of the new nation-state, nationalism has become the core of most people's identity. Mazzini, for all of his lack of a formal philosophical system, has won over Marx and Mill. The arational has won over the coherence of reason.

The nastiest form of nationalism is the ethnic variety. Europe has much experience in this kind of nationalism, for ethnic nationalism helped to cause the First World War, was a significant cause of the Second World War, and propelled local hostilities at other times in many areas. The greatest crimes of the twentieth century were underscored by ethnic nationalism, killing whole populations in the name of loyalty to one's nation. In the 1990s it became the justification for the evil of ethnic cleansing, legitimizing further crimes against humanity. Europe's history in this matter is itself a cautionary tale.

Today, ethnic nationalism is part of the growth of populism. It is the source of Hungary's and Denmark's attempts to limit refugees and immigrants from entering their states. We now see right-wing authoritarian ethnic nationalists forming political parties and taking part in parliaments in Austria, Greece, Germany, and others. Putin uses it to justify authority and the invasion of the Ukraine. Catalans use it to attempt to secede from Spain as do those Scots who want to end their association with the United Kingdom.

There is also civic nationalism, a loyalty to a civic culture which usually has a clear constitution that includes the protection of human rights. Indeed, the EU is a place where civic nationalism is attempting to be reified—loyalty to a certain kind of Europe, not a blind loyalty to whatever it is Europe is said to be.

Nationalism of the ethnic variety is not a belief that is peaceful. Ethnic nationalists believe it is legitimate to exclude people on the basis of their religion, colour or origin. These nationalists are all over Europe, growing more powerful, and they are a great danger to everything the EU stands for.

6) How do we understand human nature? Do we see ourselves as belonging to a people? Are we members of a class? What does it mean to have consciousness? Are we descended from the animal kingdom?

We humans are far more complicated than many thought us to be in 1859. Baudelaire was closer to us in his views than some others, with his sense of the contradictions and paradoxes that are part of how humanity functions. Marx and Engels thought us better than we have turned out to be. Burckhardt, in studying Renaissance Italy as a way of understanding modernity, got close to our capacity for evil as well as our creativity. Dostoyevsky, in raising the matter of consciousness, joined Baudelaire in understanding what makes modernity unique.

Clearly, we are not as kindly or decent as was thought two centuries ago. The experience of the first half of the twentieth century did away with that perception, as did the reflections of Freud and others about our nature. If we have the capacity to be godly, we have also demonstrated a zest for evil and destruction, with the capability to do so on a grand scale in modern weaponry. We know more than any other humans before us, but we are uncertain about many important matters in the realms of values and morality.

In the next generation after 1859, it was Nietzsche who became famous for the statement "God is dead." He went further: "*We have killed him*, you and I. All of us are his murderers." It was a big metaphor, god standing for certainty of belief and a *Weltanschauung* which was generally agreed upon. There were two responses to this idea, remarked Nietzsche, in several of his works. One was to fear it, to worry that we no longer have the kind of clarity that people had, let us say, in the medieval world. The second was to welcome it, to live a life of experimentation and questing, to glory in new possibilities. That dichotomy is still with us.

Among the many original things accomplished in the modern world is the creation of new symbols and myths to help us understand what is happening. Many old myths were kinetically changed and brought into consideration, including creation myths from many cultures and important ones from both the classical world and the Judeo-Christian tradition, such as the stories of Daedalus and Icarus, Apollo and Dionysus, Jacob and Joseph, and the Exodus.

Yet it was necessary to invent new ones to get at the uniqueness of the modern sensibility. Two of the most influential in modernity are inventions of those two great writers who died in April, 1616: Miguel de Cervantes and William Shakespeare.

Cervantes' Don Quixote is a character we cherish. A man of middle-age, on a scruffy plain in the middle of Spain, reads books about chivalry and decides to become a knight to help make the world a better place. He is considered mad by many, but it is a kind of divine madness. Everyone he comes into contact with is transformed in one way or another. He takes on impossible causes and is constantly faced with failure. His battle with modernity is symbolized by his fight with the windmills, which represent innovation and change. He is the first Luddite. Yet, in spite of getting beaten, being vilified, and simply often failing, he continues. He is, as we now say, quixotic, but he is a figure to whom we can relate.

Hamlet is also all of us, but in a different mode. He is faced with choices in important moral situations and doesn't know how he should act or what is the right thing to do. He vacillates, he toys with madness, he has visions, he lives inside his consciousness for a time. His uncertainty, his doubts, and his mental disturbance are completely modern. He is as close to us as some characters in Beckett or Pinter.

The early nineteenth century witnessed the creation of two of the most relevant and powerful tales about modernity, represented by the characters of Frankenstein and Faust.

Mary Shelley's Dr. Frankenstein (from *Frankenstein*, published in 1818) and the nameless Monster he creates raise many modern issues. How far should knowledge go? Are we humans morally prepared to deal with the knowledge and power associated with discovery? Dr. Frankenstein is morally deficient and, though he believes himself to be a discoverer, he is really a destroyer. He abandons his creation, a metaphor not only for a parent abandoning a child but also for the question raised by the book's epigraph from Milton when Adam pleads, "Did I request thee, Maker from my clay/ To mould Me man? Did I solicit thee/ From darkness to promote me?" Have we in modernity been abandoned by our god or the gods? If we take on the role of the Creator, are we capable of doing so in a responsible manner?

The Monster makes his own case after he learns language. "I ought to be thy Adam," he tells Frankenstein, "but I am rather the fallen angel, whom thou drivest from joy for no misdeed." What is the responsibility of the creator? What are the consequences of abandonment? It was the physicist Robert Oppenheimer who, witnessing the bomb he helped to create, quoted from the Bhagavad-Gita, "Now I am become death, the destroyer of worlds."

The subtitle of Shelley's novel is "The Modern Prometheus." Hence, she borrows from both the classical and the biblical traditions to create a story that keeps informing us about modernity and its discontents.

Goethe's *Faust* (published in full in 1832) is his modern interpretation of the Faust legend which originated in Germany in the sixteenth century. As Marshall Berman points out in his insightful analysis of Goethe's work, this Faust becomes an actor, fulfilling the later Marxian axiom that philosophers should not merely understand the world, they should transform it.

The devil, Mephistopheles, makes his bargain and Faust leaves his comfortable study and goes into the whirl of modern life. He tells Mephistopheles that he must not stop. He must continue to be an agent of transformation: "Plunge into time's whirl that dazes my sense,/Into the torrent of events!... The reeling whirl I seek, the most painful excess."

Faust becomes a builder, what Berman calls a developer, and he, like the makers of the industrial revolution, is a great agent of change. He enters into the world of magic and illusion, going past his former life of reason. He finds love, but he cannot stop and cruelly abandons Margaret to continue in the whirl. He will try to do so even as he goes blind near the end of the work.

The Faustian bargain is with us all of the time in modernity. For example, the university at which I teach was only four score years ago a piece of pastoral farmland on the outskirts of a major city. Now, it has many buildings, much concrete, and it serves over 50,000 students. We have progressed in the way Faust and modernity demand. But as Goethe and Dostoyevsky point out, modernity has a price. As we build, we also destroy. As well, there are political and diplomatic bargains which sometimes make us cringe.

Other creations, many around the year 1859, include the Crystal Palace and other buildings as symbols of the new, including the Eiffel Tower, opened in 1889 to celebrate the new France coming out of its 1789 revolution, in a city which also has late-medieval cathedrals. Many railway terminals of the nineteenth century are the modern equivalent of the old churches, ornate and celebratory.

We have encountered characters that are new, including Superfluous Man and Underground Man. Others will be created and/or become symbolic. They include the Vampire, best known in *Dracula* (1897), the circus people in poems and paintings, and the revolutionary protester on the street, asking for a voice and for change.

Female characters who wish to liberate themselves are part of modern mythology. In the last half of the nineteenth century they include Emma Bovary, whom we have encountered, and Anna Karenina, both doomed by the limits of their contexts. A character who seems to succeed is Nora, from Ibsen's *A Doll's House* (1879), who decides to leave her hypocritical and foolish husband, the play ending with her exit and the slamming of a door. But the important matter is not Nora's leaving. We need a play that gives us an understanding of what happens after the door slams. It would not be until the twentieth century, and even then only in some places that women would have real choices.

A place that is given special importance grows with the rise and influence of nationalism. It began with the response to Napoleon I's attempt to conquer and control Europe. The Russian victory in 1812 was attributed by Russians and others to the power of the land, not to the guile of generals. All national movements carry with them a myth of the land, a special place associated with the nation, something unique. It is not enough to tell a people they might have a polity; they must have it in a place that has the quality of the holy.

Later, in the twentieth century, other symbols and myths will be added. They include the Problem That has No Name, the Man Without Qualities, Invisible Man, Organization Man, and the Man who Obeys Orders.

The philosopher Albert Camus revived one myth from the classical world and invented a new metaphor to help us understand and negotiate crises in modern life. The myth is that of Sisyphus, king of Ephyra, who was cruel, avaricious, and mendacious. He violated basic rules of behavior, including those related to welcoming travelers. As a result, his punishment was to roll a large rock up a hill, only to have it escape his grasp near the top, forever repeating this frustrating task.

In his essay on Sisyphus, Camus dealt with the modern concept of the absurd. Sisyphus represents the absurdity of modern life, including being in jobs which are soul-destroying. However, Camus imagines that once he reflects on his condition, Sisyphus would reach a place of peace. Realizing the absurdity of his task, he accepts his fate. Moreover, he understands that his task, as absurd as it seems, gives him a purpose. "The struggle itself...is enough to fill a man's heart. One must imagine Sisyphus happy," says Camus. It is also a very modern thing to value the process more than the result, as Dostoyevsky contemplated before Camus.

The new metaphor is that of the plague, the title of one of Camus' novels, written to deal with the events surrounding the Second World War and to reflect on our choices and their meaning. Camus was not the first to use a medical metaphor related to health. Jonathan Swift in his letters discussed the need to "heal" society. Thomas Mann placed his classic novel *The Magic Mountain* (1924) in a sanatorium. As well, in Mann's novella, *Mario and the Magician* (1929), among the first literary works to deal with the rise of fascism, Mann's narrator tells his children that the wildly nationalist Italians have a kind of illness.

The plague comes to Oran, and Camus tells us how many people deal with it. The issues around the plague are several. First, what is our responsibility when a plague appears? Some who are not of Oran tell the narrator it is not their affair. They are in quarantine and cannot leave, but they disown responsibility. Others just do their job—including our doctor-narrator who tries to comfort and heal, and a local clerk who assists—and are quiet, simple heroes. There are those who welcome the plague because they can benefit from it. Some, including a local priest, understand it as a punishment.

Camus also notes that the plague, when it comes, is arbitrary. Some die, not because of any reason, simply because they are victims of the microbe. Moreover, the plague recognizes neither wealth nor character nor status. It is absurd. Life itself is fragile.

The plague represents the evil that appears in human affairs regularly, including the Nazi occupation of France in the Second World War. Camus tells us that in modernity something like the plague regularly appears and we have profound moral questions to reflect upon. When it does appear, not through any choice made by those affected by it, the narrator insists that we must respond to counter it, even if that too is sometimes futile. One must "refuse to give way to the pestilence." Moreover, though eventually the plague may recede and normality may return, reflects the narrator, "this joy was always under threat."

Many of those who celebrate the end of the plague were

> unaware of something that one can read in books, which is that the plague bacillus never dies or vanishes entirely, that it can remain dormant for dozens of years in furniture or clothing, that it waits patiently in bedrooms, cellars, trunks, handkerchiefs and old paper, and that perhaps the day will come when, for the instruction or misfortune of mankind, the plague will rouse its rats and send them to die in some well-contented city.[157]

In Modernity one must be constantly on guard against those who represent the plague.

Sometimes it is useful to go back to old myths and revive them, as Camus did with Sisyphus. In Ovid there are two stories which are sometimes overlooked, but have special resonance in modern times. They are the tales of Baucis and Philemon, and that of Erysichthon. In the *Metamorphoses*, how Ovid places his myths is significant. These two are together, first Baucis and Philemon, followed by Erysichthon.

Baucis and Philemon are two good old married people, very poor, but yet cheerful about their life together. They are visited by Jupiter and Mercury, gods disguised as ordinary men, who had sought rest in many houses and been turned away. Baucis and Philemon do the right thing according to the customs of their culture. They welcome these unknown visitors who knock on their door, feed them from their meagre store of food, offer wine, and light a fire for warmth. They even attempt to kill their pet goose to nourish the visitors.

The two gods stop at that point and reveal their identities. Those who turned them away will experience a flood which will destroy their homes. However, the poor home of the couple will be transformed into a temple as part of their reward. In addition, Jupiter and Mercury tell the couple, "You are good people, worthy of each other,/ Good man, good wife—ask us for any favour,/ and you shall have it." The couple ask for some time to reflect and talk about it. Then, the opposite of the greedy Midas, they ask to be the guardians of the temple until it is their time to die. "May one hour take us both away," they ask, may neither of us have to bury the other. Their wish was granted and as a sign they turn into trees, an oak and a linden, that becomes one.

---

157 Camus, *The Plague* (London: Penguin, 2002), 237–38.

The characters of Baucis and Philemon will play a role in Goethe's tale of Faust. Near the end of the work, Faust has become a builder and a modern agent of transformation. The couple had lived in a kind of Eden. Philemon tells a Wanderer, who represents the gods, "You see gardens now and acres/ Of a paradisiac sight."

But they are in the way of what Faust and others would call progress. The builders are moving along: "Clever masters' daring slaves/ Toiled till dams and trenches spread,/ Pruned the power of the waves/ To be masters in their stead." The new industrial world, in control of nature, is relentlessly spreading, "offering men a new existence."

Faust needs to get the couple and their pastoral world out of the way. Baucis speaks,

> Daily they would vainly storm,
> Pick and shovel, stroke for stroke;
> Where the flames would nightly swarm,
> Was a dam when we awoke.
> Human sacrifices bled,
> Tortured yells would pierce the night,
> And where blazes seaward sped
> A canal would greet the light.
> He is godless, covets our
> Cottage and our wooded fringe;
> As the neighbor swells with power,
> We should crouch, and we should cringe.

The world of Baucis and Philemon is at risk. And in this new world of railways, dams, canals, and cities, there will be human sacrifices in the service of modernity.

Erysichthon is as evil as Baucis and Philemon are good. He was the king of Thessaly, a monarch, as Ovid relates the story, who "scorned the gods." In his arrogance he attacked a sacred grove of the goddess Ceres, the Roman deity of agriculture, fertility, and the earth. In the grove there was an oak tree which was also sacred. Erysichthon ordered his slaves to fell the tree, but they recoiled. Hence, he took an axe and chopped it down, killing someone who tried to stop him. A nymph of Ceres, who he also killed, put a curse on him as she was dying.

Ceres sent a messenger to Famine, who flew to Erysichthon, breathed into his body, and "planted hunger in his hollow veins." He dreamed and thought only of food, but he could not be satisfied. Everything was consumed. He then sold his daughter, who had the capacity to change her form time and again, in order to get more food: "Till finally there was nothing, nothing, only/ His own flesh for his greedy teeth to seize,/ To gnaw on, and the wretch consumed his body/ Feeding upon a shrinking self."

We need stories. Sometimes we need to invent stories to help understand the human condition in our time, sometimes we can use the stories of the past and apply them to the present.

We also still have plagues which appear arbitrarily in our world. One which is getting worse each year is that of the climate. Are we ignoring the plague until the time when it might simply overwhelm us? Are we emulating Erysichthon in our arrogance of power and in the destruction of the very environment we need to sustain us? Are we devouring everything to the point where we will destroy our own selves? Do we ignore the warnings of nature and the gods at our peril?

Modernity is a place of paradox and contradiction. We are powerful and vulnerable at the same time. We are godly and we are evil like Erysichthon, generous and avaricious, charitable and cruel. Above all, we need to reflect on the issue raised in the works of Marx, Engels, Darwin, and others: how do we extend the benefits of modernity to as many as possible in a just manner? Our 1859 group coped with the beginnings of modernity. The first half of the twentieth century was mainly destructive and fraught with danger. Our time is faced with the ultimate question, that of survival. We have the capacity to destroy the planet through weaponry. We have the capacity to destroy it through greed and ignorance. We also have the possibility of making a better, safer, sustainable world. As Camus would tell us, whether we like it or not, whether we act or not, we are always making choices.

It is Italo Calvino, following Camus, who in his *Invisible Cities* (1972), borrowing from Dante, captured the essence of our present time:

> The inferno of the living is not something that will be; if there is one, it is what is already here, the inferno where we live every day, that we form by being together. There are two ways to escape suffering it. The first is easy for many: accept the inferno and become such a part of it that you can no longer see it. The second is risky and demands constant vigilance and apprehension: seek and learn to recognize who and what, in the midst of the inferno, are not inferno, then make them endure, give them space.[158]

What we do know is that change is the law of our world. Another of Ovid's characters, Pythagoras, also tells us about modernity near the end of the *Metamorphoses*:

---

[158] Italo Calvino, *Invisible Cities*, trans. William Weaver (New York: Harcourt Brace Jovanovich, 1974), 165.

All things are fluent; every image forms,
Wandering through change. Time is itself a river
In constant movement, and the hours flow by
Like water, wave on wave, pursued, pursuing,
Forever fugitive, forever new....
What we have been,
What we are now, we shall not be tomorrow.

Index

aesthetics   8, 79, 136, 180, 210
Alexander II   61
Alighieri, Dante   84, 88, 90, 219
anarchism   53, 112, 171, 173, 175–177
anarchists   53, 93, 171–173, 175, 177–179, 186
ancien régime   1, 47, 48, 51, 55, 125
Aristotle   97, 105
Association   51, 54, 117, 143, 177, 204
Austria   42, 185, 212
Austrian Empire   *See* Austria
authoritarianism   93, 98, 102, 110, 179

Bakunin, Mikhail   53, 74, 143, 171, 176–179
Barcelona   112–114, 130, 133–136, 138–142, 149
Baudelaire, Charles   3, 7–12, 14–25, 27–32, 38, 40, 85f., 88, 97f., 113, 122, 127, 166, 183f., 209f., 212
Becoming   9, 11, 17, 21, 44, 63, 73, 97, 108, 117, 122, 130, 140, 148, 153, 163f., 177, 199, 211
Being   1, 4f., 7–9, 16, 20f., 26, 30, 32f., 41–46, 49–51, 54–57, 61, 63f., 68, 76, 78, 80f., 83f., 92, 96f., 100f., 105, 107, 110, 114, 127, 132, 135, 142, 146f., 150f., 153, 157, 160f., 163, 165, 171, 175f., 178, 180, 187, 192, 194, 196, 201, 204, 209, 213, 216, 219
Benjamin, Walter   12, 34
Bentham, Jeremy   94f.
Berlin   43, 78, 92, 130, 156, 176, 182, 210
bolshevism   98, 110
Bonaparte, Napoleon (Napoleon I)   6, 18, 41, 43, 60, 122–125, 129, 131, 134, 157, 160–162, 170, 183, 215
Bonaparte, Louis Napoleon (Napoleon III)   6, 18, 122, 125, 129, 131, 134, 158, 160–162, 170, 183
Boudin, Eugène   12
bourgeois/ie (class, revolution, society)   1, 9f., 12, 20f., 24, 30, 34, 40, 59, 76, 86, 89, 104, 119f., 123, 126f., 132, 135f., 151–156, 158–161, 163f., 177, 179, 181, 183
Britain   42, 55, 59f., 74, 93–95, 99, 102, 141, 166f., 180–182, 184, 194f., 200, 202, 205
Bukharin, Nikolai   71
Burckhardt, Jacob   1, 7, 78–92, 151, 158, 210, 212
Burke, Edmund   70, 93, 104

Caillebotte, Gustave   13f., 35
Calvino, Italo   219
Camus, Albert   17, 185, 216f., 219
capitalism   1, 101, 112, 165f., 168, 174, 183, 209
capitalist   21, 59, 100, 111, 141, 164–166
Carlyle, Thomas   46, 180
Cassatt, Mary   3, 16
Catherine the Great   58
Cerdà, Ildefons   112–114, 134–141, 149, 200, 209
Christianity   48, 54, 86, 91f., 104, 148
Church, the   2, 18, 21, 48, 50, 52, 59, 82, 86f, 93, 102, 104, 109, 160, 169–171, 203, 210, 215
Cicero   4
civic nationalism   *See* nationalism
colonialism   99, 165
communism   53, 93, 110, 112, 147, 156, 172f., 206
communist   55, 93, 143, 150, 155f, 167f., 177, 179, *See* communism
Comte, Auguste   5, 44, 139
Condorcet, Marquis de   5–7, 17, 49, 168
Congress of Vienna   6, 42f., 60, 93
conservatism   93, 104, 124, 212
Constantin Guys   8, 12
Crimean War   61, 64

dandy   8, 15, 29f., 33
Darwin, Charles   2f., 9, 44, 47, 97, 139, 150, 187–208, 210, 219

Darwin, Erasmus 87–88
Daumier, Honoré-Victorin 12, 29, 122, 127, 162
decentralization 173
Delacroix, Eugene 8, 30
devil 26f., 185, 214
Dickens, Charles 754, 95f., 122, 188
Disraeli, Benjamin 74, 121
Dostoyevsky, Fyodor Mikhailovich 17, 58, 63, 97, 181–186, 210–212, 215f.
Dumas, Alexandre 3

education 5, 51–53, 56f., 94–97, 104, 106, 108–110, 155, 175, 210f.
educational system See education
Eliot, George 3, 79, 188
empire 41, 76, 81, 93
Engels, Friedrich 9, 20, 86, 104, 113f., 118–122, 127, 143, 145, 147–156, 161, 163–168, 173, 177, 193, 197, 206f., 210–212, 219
England 1, 10, 46f., 95, 99, 102, 110, 113f., 118–121, 142f., 145, 156, 177, 187, 200
Enlightenment 1, 5f., 26, 41, 49, 59, 80, 91, 147, 163
ethnic nationalism See nationalism
Europe 1, 3, 6, 10, 14, 41–43, 46, 49f., 52f., 55–58, 60, 64, 76, 78, 80–86, 91–95, 104f., 109, 111, 114, 124, 128f., 132f., 139, 141f., 149, 152, 156f., 165–167, 175, 181–183, 186, 204f., 209–212, 215
European Union 47, 111, 209, 211f

fascism 55, 57, 98, 110, 216
fascists 109, 124, 162, 167
Faust 26, 70, 120, 147–149, 214f., 218
federalism 49
feudalism 70, 151
flâneur 11–15, 19f., 22f., 26f., 30, 99, 132, 184
Flaubert, Gustave 8, 19, 36, 38, 58, 127
France 6, 9f., 18, 42, 46, 52–60, 70, 73, 80f., 95, 104, 109, 122–125, 127–129, 131, 135, 141f., 150, 156–160, 162, 166f., 171, 175, 177, 181f., 210f., 215f.

Franco-Prussian War 18, 55, 122, 160, 170
French See France
French history See France
French Revolution 1, 5f., 41, 46, 59, 93, 128, 132, 151, 157f., 169, 179
Freud, Sigmund 44, 128, 156, 168, 207, 213

Garnier, Charles 131f., 142
Gaskell, Elizabeth 3, 74, 121
Gaudi, Antonio 142
Gavarni, Paul 12
German See Germany
Germans See Germany
Germany 10, 43, 55f., 58, 119, 156, 162, 166, 182, 189, 212, 214
God 4, 8, 26f., 47–49, 54, 64f., 91, 146, 148, 178, 191–194, 201–203, 213
Goethe, Johann Wolfgang von 70, 120, 147f., 214f., 218
Gothic 141, 153, 181
government 45–47, 50, 53f., 58, 61, 93, 98, 101f., 108–110, 112, 114f., 117, 122–125, 127, 131, 134, 136, 139, 157, 159–161, 166, 172f., 175f.
Great War 56

Haussmann, Georges-Eugène 18, 21, 25, 35, 122–125, 128f., 135, 139, 149
Hegel, Georg Wilhelm Friedrich 17, 41, 44, 48, 55, 78, 80, 144f., 148, 150f., 158, 163, 172, 177
Historicism 1, 44
history 2–4, 8f., 12, 18, 21, 23, 31, 41–45, 47, 50, 52, 54, 59–61, 67, 69, 73, 78–81, 83, 87f., 90, 92, 94, 98, 102, 109, 117, 139, 144–147, 149–152, 155, 157f., 161, 163–165, 181, 184, 187, 189, 191, 196, 199f., 202, 204, 206, 210, 212
Hobsbawm, Eric 1, 52, 109, 186, 210
Hugo, Victor 18f., 58, 167
humanists 4, 85–87, 91
Huxley, Thomas Henry 193, 203–205

imperialism 99
impressionism See impressionist

impressionist 13, 33
impressionists 33, 35, 181
Industrial Revolution 1, 6, 35, 59, 74, 114, 116, 120, 142, 145f., 151, 158, 165, 171, 175, 179f., 191f., 214
Italian *See* Italy
Italian Renaissance *See* Renaissance
Italians *See* Italy
Italy 1, 7, 41f., 44–50, 52–55, 57, 78–92, 151, 162, 169, 177, 182, 212

July Revolution of 1830 10

Kant, Immanuel 5, 7
Karatayev, Vassily 63
Kierkegaard, Søren 17

Lenin, Vladimir Ilyich Ulyanov 55, 71, 165
liberalism 50, 55, 93, 98, 104, 107, 110, 112, 169f., 186, 203
liberty 2, 46, 49, 51–53, 57, 98, 100f., 104f., 107f., 110f., 134, 173, 175–177, 183
Lombardy 42, 44, 49
London 9, 11, 30, 46, 52, 56, 67, 96, 114f., 119, 122, 125, 129f., 133, 143, 149, 157, 172, 176, 178, 180, 182, 188, 204, 206, 217
Louis-Philippe 10
Louis XIV 81, 128, 131, 170
Lyell 187, 190–192, 194, 201–203, 206

Machiavelli, Niccolò 5, 83f., 90–92
Machiavellian *See* Machiavelli
Malthus, Thomas 191f., 195
Manchester 11, 112, 114–121, 123, 126, 130, 133, 142f., 151
Manet, Édouard 7f., 14, 27, 29–40, 86, 122, 210
Marx, Karl 6f., 9, 11, 17, 20–22, 38, 44, 47, 49, 53, 55f., 64, 73, 76, 86, 91, 104, 113, 119f., 127, 138–140, 143–159, 161, 163–168, 171–173, 176–180, 192f., 197, 206f., 209–212, 219
Marxist 57, 161, 178
Marxists 143, 156, 177, 179, 206f.

Mazzini, Giuseppe 2, 9, 41–43, 45–57, 65, 93, 143, 169, 175, 177, 209, 212
Milan 82f., 89
Mill, John Stuart 2f., 9, 38, 46f., 49, 51–53, 56f., 59, 86, 93–111, 118, 146, 157, 165–168, 198, 202f., 209–212
Milton, John 4, 89, 214
modern 1f., 4–11, 14f., 17f., 21, 23f., 27, 30, 32f., 35, 41, 44, 48, 52–55, 57, 61, 66, 70f., 80–92, 94, 104, 109, 112f., 120, 122–124, 127, 129, 131, 135, 142, 145, 149, 151–154, 157, 164, 166, 169, 171f., 179–182, 184f., 189, 209–211, 213–218
modernity 1–3, 7–9, 11f., 16f., 19, 24f., 27, 31, 33, 35, 37, 39f., 44, 47, 55, 70, 79–81, 86–88, 90, 92, 97–99, 101f., 104f., 108–112, 114, 118, 123, 129f., 132, 142, 148f., 157, 166f., 169, 171, 175, 179–181, 183f., 186, 200, 208, 210, 212–216, 218f.
Monet, Claude 14, 32, 35f.
Monlau, Pedro Felipe 134
Moscow 59, 64, 69, 119, 144
Mussolini, Benito 48, 55

nationalism 2, 41, 43–45, 47f., 50, 52, 54–57, 77, 93, 156, 166f., 169f., 175, 209, 211f., 215
Nicholas II 60
Nietzsche, Friedrich 10, 17, 92, 213
nihilism 68, 70, 112
nihilist *See* nihilism

Orwell, George 99, 179
Ottoman Empire 42, 60, 76
Ovid 4, 18, 22, 217–219
Owen, Richard 202, 204f.

Paris 3, 8, 10–16, 18f., 21f., 24f., 27, 30, 33, 35f., 38, 112, 114, 122–135, 137, 139, 142, 157, 161, 164, 182f.
Parisian lifestyle *See* Paris
Patriotic War of 1812 60
Peter the Great 58
Petrarch, Francesco 4, 88

philosophy 2, 11, 17, 40, 47f., 51, 64, 70, 78, 85, 95, 102, 107, 120, 144–146, 148, 150f., 158, 170, 176, 194, 209
Piedmont 44f., 47, 49
Pissarro, Camille 33
Pius IX 47, 55, 169–171
Portugal 42, 55, 162
proletariat 119, 152–156, 159–161, 163–165, 167, 178
Prussia 42, 185

Railways 35
Ranke, Leopold von 43, 78, 80, 92, 210
Realists 10, 22
Renaissance 1, 4, 35, 37, 40, 59, 78–82, 84–92, 131, 151, 158, 181, 212
Renoir, Pierre-Auguste 15
Roman Catholic Church 48, 103, 169–171, *See* Rome
Roman Empire 42, 45, 50, 81
romanticism 10, 22, 112
Romantics 88, 90
Rome 18, 45, 47, 87, 98, 169–171, 200
Rousseau, Jean-Jacques 5, 51, 107, 139, 156, 173
Russia 1, 41f., 53, 55–67, 69–76, 93, 102, 140, 168, 177, 179, 182–184, 186, 209f.
Russian literature *See* Russia
Russian Orthodox Church 59
Russian society *See* Russia
Russians *See* Russia

Saint-Simon, Henri de 5–7, 11, 46, 127, 139
Sand, George 3, 38, 203
scientific socialism *See* socialism
Second Empire of France *See* France
Second World War 167, 209, 211f., 216
Sedgwick, Adam 202
Shelley, Mary 95, 189, 214

Sicily 44f., 49
Slavophile 41, 58f., 182, 185
socialism 1, 21, 54f., 57, 93, 112, 147, 155f., 163, 165f., 170f., 174, 177f., 203, 211
socialist 46, 53, 56, 115, 143, 147, 155, 159, 166, 175–177, 206
sociology 30, 120, 139
Spain 41f., 45, 55, 114, 133, 135, 141f., 162, 177, 179, 212f.
Spencer, Herbert 44, 139
St. Petersburg 58f., 67, 130
sublime 10f., 24, 50

Taylor, Harriet 97
Tennyson, Alfred 192, 197
the West 1f., 4, 31, 40, 56, 58–60, 63f., 94, 105, 110, 115, 152, 157, 184–186, 188, 190
theology 2, 48, 190, 210
Tocqueville, Alexis de 82, 99, 105, 110, 118, 127f.
Trotsky, Leon 71
Turgenev, Ivan 1, 47, 49, 56–59, 61, 63–68, 70–76, 97, 176f., 211

United States 41, 50, 56, 104, 115, 167f., 185, 203

Venetia 44, 49
Venice 66, 83
Vico, Giambattista 44
Vienna 66, 130

Weber, Max 110
Wilberforce, Samuel 180, 204f.
Woolf, Virginia 165

Young Italy 45

www.ingramcontent.com/pod-product-compliance
Lightning Source LLC
Chambersburg PA
CBHW050523170426
43201CB00013B/2061